Advanced Technologies and Societal Change

This series covers monographs, both authored and edited, conference proceedings and novel engineering literature related to technology enabled solutions in the area of Humanitarian and Philanthropic empowerment. The series includes sustainable humanitarian research outcomes, engineering innovations, material related to sustainable and lasting impact on health related challenges, technology enabled solutions to fight disasters, improve quality of life and underserved community solutions broadly. Impactful solutions fit to be scaled, research socially fit to be adopted and focused communities with rehabilitation related technological outcomes get a place in this series. The series also publishes proceedings from reputed engineering and technology conferences related to solar, water, electricity, green energy, social technological implications and agricultural solutions apart from humanitarian technology and human centric community based solutions.

Major areas of submission/contribution into this series include, but not limited to: Humanitarian solutions enabled by green technologies, medical technology, photonics technology, artificial intelligence and machine learning approaches, IOT based solutions, smart manufacturing solutions, smart industrial electronics, smart hospitals, robotics enabled engineering solutions, spectroscopy based solutions and sensor technology, smart villages, smart agriculture, any other technology fulfilling Humanitarian cause and low cost solutions to improve quality of life.

More information about this series at https://link.springer.com/bookseries/10038

Shyamapada Mukherjee ·
Naresh Babu Muppalaneni · Sukriti Bhattacharya ·
Ashok Kumar Pradhan
Editors

Intelligent Systems for Social Good

Theory and Practice

Springer

Editors
Shyamapada Mukherjee
Department of Computer Science
and Engineering
National Institute of Technology Silchar
Silchar, Assam, India

Naresh Babu Muppalaneni
Department of Computer Science
and Engineering
National Institute of Technology Silchar
Silchar, Assam, India

Sukriti Bhattacharya
Luxembourg Institute of Science
and Technology
Belvaux, Luxembourg

Ashok Kumar Pradhan
Department of Computer Science
Engineering
SRM University
Guntur, Andhra Pradesh, India

ISSN 2191-6853 ISSN 2191-6861 (electronic)
Advanced Technologies and Societal Change
ISBN 978-981-19-0769-2 ISBN 978-981-19-0770-8 (eBook)
https://doi.org/10.1007/978-981-19-0770-8

This Springer imprint is published by the registered company Springer Nature Singapore Pte Ltd.
The registered company address is: 152 Beach Road, #21-01/04 Gateway East, Singapore 189721, Singapore

Preface

There has been a dramatically increasing interest in Artificial Intelligence (AI) in recent years. AI has been successfully applied to societal challenge problems, and it has a great potential to provide tremendous social good in the future. AI has the potential to help tackle some of the world's most challenging social problems. AI capabilities that could be used to benefit society fall into three major categories: computer vision, natural language processing, and speech and audio processing. While the social impact of AI is potentially very large, certain bottlenecks must be overcome if even some of that potential are to be realized. The most significant bottlenecks are data accessibility, a shortage of talent to develop AI solutions, and "last-mile" implementation challenges. Intelligent Systems can present an opportunity to build better tools and solutions to help address some of the world's most pressing challenges and deliver positive social impact. The content of this book is useful for young researchers and application-domain experts to identify key opportunities for future intelligent system applications targeted toward social good. This book highlights the connections between various recent technologies: Artificial Intelligence (AI), Machine Learning, and Internet of Things (IoT). It presents the application of these technologies to solve various societal problems related to healthcare, agriculture, green environment, renewable energies, smart cities, etc. This book is a contributed volume aims at collecting the contributions, which integrate Artificial Intelligence, Machine Learning, Deep Learning, and Federated learning with IoT, to give the reader an up-to-date picture of the state-of-the-art on connection between computational intelligence, machine learning, and IoT.

Silchar, India — Dr. Shyamapada Mukherjee
Silchar, India — Dr. Naresh Babu Muppalaneni
Belvaux, Luxembourg — Dr. Sukriti Bhattacharya
Guntur, India — Dr. Ashok Kumar Pradhan

Introduction

Intelligent systems are mainly high-tech machines that can perceive and respond to their surroundings. It is a machine with an enclosed, Internet-connected device, which can collect and analyze information as well as communicate with other devices also. Intelligent gadgets and interconnected collections of devices, such as network system and other more significant systems, are included in intelligent systems. Similarly, complicated Artificial Intelligence-based software process, examples such as chatbots, expert systems, and other types of software system can also be included in intelligent systems as well. Intelligent systems can be found in a variety of places, including digital televisions system, Point-of-Sale (POS) terminals, traffic lights, smart meters, vehicles, digital signaling, and flight controls etc. Built-in intelligence is an important part of the emerging Internet of Things (IoT), where nearly every single thing as it may be given a distinctive identification and the capability to automatically transmit data over a network system without the requirement for human-to-human or human-to-computer contact.

Intelligent System (IS) is a term that refers to a system that blends intelligence into machine-assisted applications. In creating intelligent systems, several kinds of machine learning approaches are there, for example, supervised, unsupervised, and reinforcement learning, can be modeled. Intelligent systems can also do complicated automated activities that the standard computing paradigm can't handle. Intelligent techniques used in Intelligent System Design have produced in a variety of diagnostic, robotics, and engineering systems. As a result, it is a technology created to execute activities that require intelligence. The creation of such a system is now an information technology engineering field that necessitates the use of effective methods and tools. The accurate characterization of an intelligent system is difficult since it is reliant on terms linked to cognition, is a field that is still poorly understood and subject to various interpretations.

Intelligent system is unreal, which is constructed from Reasoning, Learning, Problem Solving, Perception, and Linguistic Intelligence in Fig. 1. we mention it properly. Reasoning is a series of procedures that allow us to make judgments, make decisions, and make predictions. Learning is the process of obtaining information or skill through study, practice, being taught, or experience. Problem solving is the

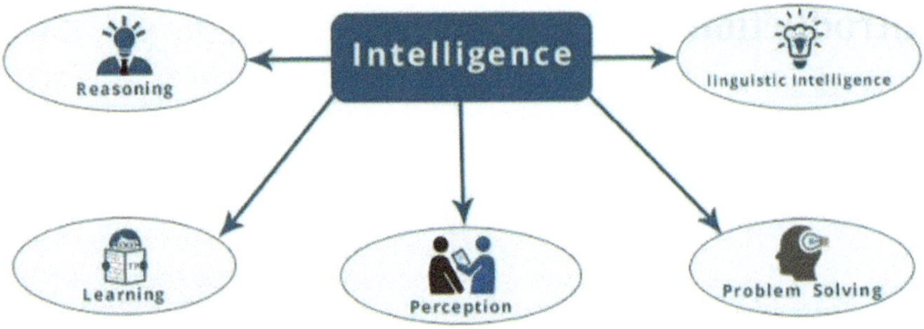

Fig. 1 Intelligence components [1]

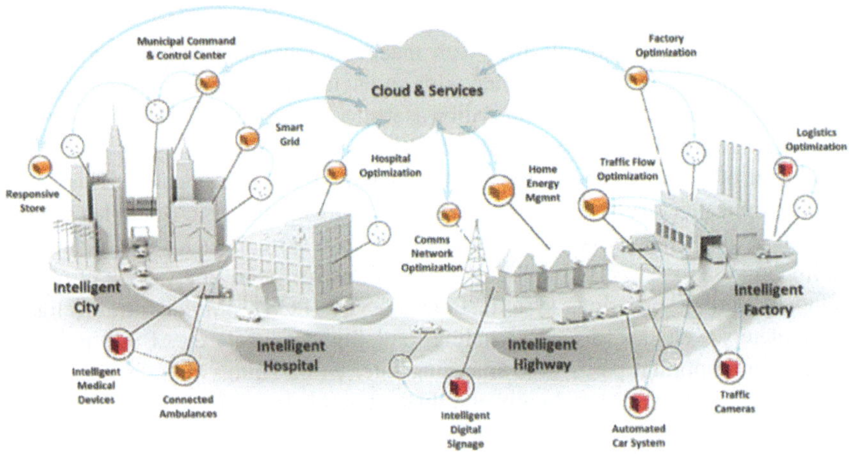

Fig. 2 IoT intelligent system framework [3]

process of perceiving and attempting to reach the desired solution from a current circumstance by following a path that is hindered by known or unknown obstacles. Perception is the process of collection, interpretation, selection, and organization of sensory data. Linguistic intelligence is one's power to use, comprehend, speak, and write the oral and written language [2].

The term "intelligent system" is used to describe a set of cognitive capacities such as perception, action control, interaction, and deliberation. Different sorts of systems can be distinguished by the existence or lack of such abilities. Figure 2 depicts some of the most common forms of intelligent systems in a framework. This is the brief overview of various system architecture with intelligent IoT system.

The Internet of Things (IoT) is a network of interconnected computing devices, digital and mechanical ingredient, humans or animals, and objects that can perceive, gather, and transport data over the internet without the requirement of human interruption. An IoT is created by web-enabled smart components that collect, transmit, and

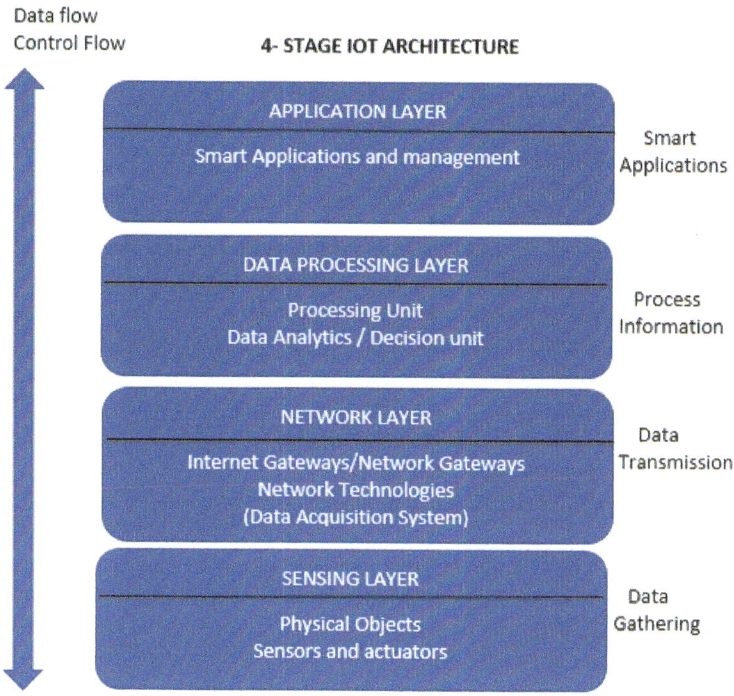

Fig. 3 IoT architecture

function on data from their neighbors using embedded systems, for example, processors, sensors and communication hardware system. IoT devices can exchange sensor data delivered to the cloud system for analysis or processed locally by connecting to an IoT gateway or other edge device. Artificial Intelligence (AI) and machine learning are helpful for IoT to prepare information gathering processes dynamic and straightforward. Every day, new technologies arise, or existing ones are updated. The recent developments are the 5G network system. In IoT systems and applications, 5G expect to play a critical role. Its high frequency and bandwidth attract researchers' notice and stimulate their interest in possible security and privacy problems. However, the short-wavelength requires a shift in infrastructure, necessitating more base stations to cover the equal region as other wireless technologies. Fake base stations, for example, pose a more significant hazard under this new structure. Recognize the security issues and the potential solutions are essential.

Architecture of IoT

There are mainly four stages available. Figure 3 represents these four stages properly.

- Stage 1: Sensors and Actuators: Sensors and actuators are the components for sending, receiving, and processing the data all over the network system. These components are merged with cable wire or maybe a wireless system. RFID, GPS, electrochemical, gyroscope, and other technologies are also included in these technologies. Local Area Network or Personal Area Network can be helpful to connect sensors and actuators altogether.
- Stage 2: Data Acquisition System: This system is mainly used to collect new/fresh values from the sensor side and converts these values from analog to digital format. Then combine and prepare the data before transmitting it to the upcoming level of processing with the help of an Internet gateway, wireless wide-area networks (for example, Wi-Fi system or Cellular network system), or wired wide-area networks.
- Stage 3: Preprocessing units: Data analytics: After the IoT data have been digitized and entirety, it will need to be processed to reduce the data amount even further earlier on being address to a data center or cloud. As part of the pre-processing, the edge device may execute some analytics.
- Stage 4: Smart application and management: Data center or cloud: The Management Services group includes the Data Center or Cloud, which processes data through analytics, device management, and security controls. Apart from security manages and device administration, the cloud system also sends information to end-user, including retail service, health systems, emergency cases, environmental issues, energy resources, and others—Fig. 3 shows these stages very clearly [4].

Security and Privacy Issues in IoT

Likewise, other technology system, the IoT has both advantages as well as disadvantages. Here, proponents of technology system and manufacturers who manufacture the IoT components promote the IoTs as a way to create our regular life better and more accessible by connecting billions of "smart" IoT components (for examples, Smart TVs, Refrigerators, Ovens, Cameras, Smart Doorbells, Smart Healthcare and Acting Tracking, and Smart Police Surveillance and Traffic Systems, etc.). Because of user privacy and data security risks with IoT devices, IT security professionals admit it is unessential and dangerous. As reported by the findings of a recent Altman Vilandrie & Company survey of 397 IT leaders from 19 industrial companies, nearly half of all U.S. organizations that employ IoT components were breached in 2017.

- Security system: Since the IoTs are different from conventional electronic and computing devices, it is more defenseless to security threats in numerous ways. Many IoT components are created to be set in huge numbers. Sensors are the great instance of it. Generally, an IoT deployment comprises of a group of comparable or nearly identical equipment with similar properties. This uniformity amplifies the impact of any security flaw that could harm a huge number of them. The man in the middle attack, in which a third party hijacks a communication channel with the goal of faking the identities of the real nodes include in the network system exchange, is one of the most common assaults in the IoT system. Because the opponent does not require to know the alleged victim's identification, this attack efficiently creates the banking server acknowledge the transaction as a real occurrence.

- Privacy system: The functionality of the IoT is deciding on how much successfully it can favor people's secret preferences. It is critical to feel that client secret rights and favor are troublesome in ensuring users' confidence and self-assurance in IoTs, connected devices, and associated services. Lots of action are being done to conform that IoT redefines privacy problems like raised spying and tracking.

Existing Solution

Secure the IoT Network—By deploying typical endpoint security system forms like antivirus, anti-malware, firewalls, and trespass prohibition and detection process, we can defend and safe the networking system connecting IoT components to back-end systems on the Internet connecting system. Authenticate the IoT devices—introduce numerous client management forms for a individual IoT component and implement strong authentication architecture for example, two-factor authentication system, digital certificates, and biometrics system to allow clients to authenticate IoT devices. Use IoT PKI Security Methods—use IoT public key infrastructure security mechanisms like X.509 digital certificates, cryptographic keys, and life-cycle features with public or private key generation, distribution, maintenance, and rescission to provide a secure connection within an IoT device and application [5]. Use IoT Security Analytics—it can identify IoT-specific threats and trespass that traditional network security solutions like firewalls can't track out.

Intelligent IoT Solutions for Smart eHealth

Figure 4 shows the classification by which we can use intelligent IoT system in e-Health center and help us the smart health facility system.

Intelligent Green Computing in the IoT System

To fulfil the ever-increasing need for IoT, cloud computing services are deployed. Data centers are increasingly becoming one of the top energy consumers in order to support the IoT paradigm's architecture. As more innovations arise and technology adopts new practices, the demand for energy will rise in the future, culminating in the adoption of green computing. Green computing solutions help IoT devices consume

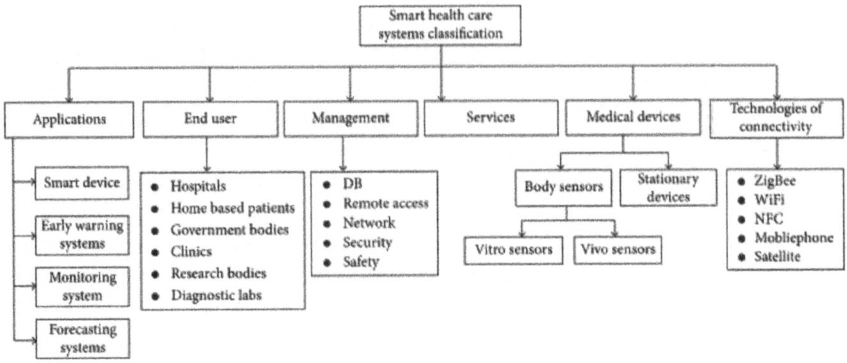

Fig. 4 Classification of smart healthcare system using intelligent IoT system [6]

less energy while maintaining their performance. This study examines a variety of aspects of green computing for IoT computing, including key principles, difficulties, and solutions.

Intelligent Industrial IoT

Manufacturing companies must increasingly embrace the industrial Internet of Things (IIoT) in order to remain competitive. This means that establishing a digital organization is a must, as Bain & Company believes that the IIoT market would be worth $200 billion by 2021. The usage of smart sensors and actuators to improve constructing in industrial sector and processes it smoothly is called as Industrial Internet of Things (IIoT). IIoT, acquainted as the industrial internet system or Industry 4.0, takes benefits of the data that "dumb machines" have built in industrial settings for years with the help of the power of smart machines and real-time analytics system. There are various Industrial IoT Applications present, for example—ABB: Smart robotics, Amazon: Reinventing warehousing, Cisco IoT, Bosch: Track and trace innovator etc.

Intelligent IoT Communication and Network Virtualization with Intelligent Edge/Cloud Computing

The Internet of Things (IoT) is considered as one of the most disruptive technologies of the future, and it has received a lot of research interest in recent years. IoT devices are small sensing or actuating devices attached to everyday objects that can send and receive data and orders. To overcome the restricted resources of IoT devices, cloud computing technology provides massive computational and storage power over the

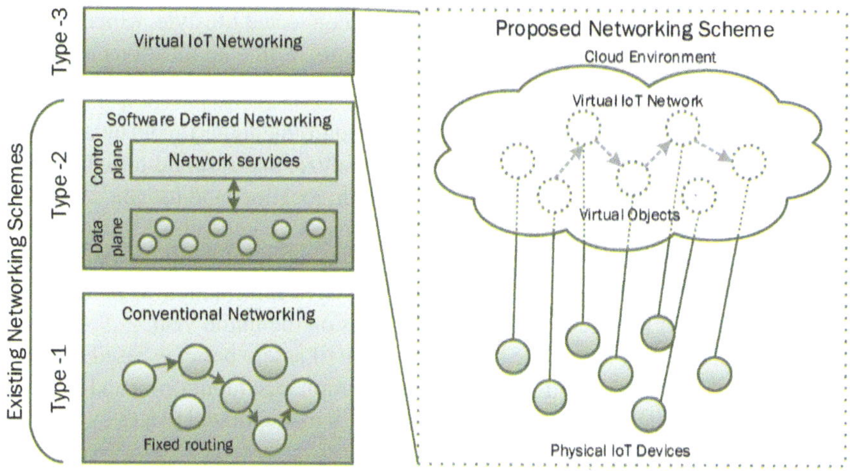

Fig. 5 Conceptual view of a virtualized IoT network [7]

Internet. Many studies on IoT device virtualization in the cloud to ease remote access and control have been undertaken. IoT devices will be accessed in the future by virtual objects that relate to them. In the cyber world, there must be a network of virtual things, just as a network of physical items. There is a way for establishing a virtual network among linked IoT devices from various domains using virtual objects in the cloud. By providing a dynamic end-to-end connection between IoT devices, this will promote resource sharing and the rapid creation of varied applications on top of the virtualization layer. The notion of IoT network virtualization is based on NFV, which is used in association with SDN and cloud computing. NFV allows for the rapid development of real-world device functionality in the form of a software package and its easy deployment in the cloud. Furthermore, through NFV, these virtual devices may be quickly updated, improved, and replicated Fig. 5 depicts a conceptual picture of the establishment of a virtual IoT network among virtual IoT devices (virtual objects). It also includes a synopsis of the evolution of network kinds. A type-1 network is similar to traditional physical networks in which routing information is kept at intermediate routers. Individual networking device maintenance becomes a tiresome process as the network develops in size. Type-2 networks, i.e., SDN, were introduced to address these concerns. Data and control plans are segregated in SDN. Virtualized IoT networks are Type-3 networks in which the network of linked IoT devices is maintained in the cloud via correspondence virtual objects.

Intelligent IoT Applications

The IoT is no longer a secret hype. It is a technology system that has quietly acquired traction and is now altering our future invisibly. The IoT is the outcome of humanity's curiosity and desire to live a more convenient and connected lifestyle, which reduces work and eliminates the possibility of human errors. Some applications related to intelligent IoT are:

1. Smart manufacturing: Predict maintenance requirements and avoid unscheduled downtime. In production settings, detect, analyze, and respond to IoT signals. Allow for real-time visibility into asset health, location, and usage.
2. Service monitoring: Connect IoT devices to service centers for proactive and quick problem solutions. Enhance user service by allow of agents to send instructions directly to assets to handle problems. Give information to field service experts so that they can address the problems on the initial visit.
3. Workplace safety: Automate the enforcement of safety policies based on rules. It reduces response time for accidents and compliance monitoring and prevents harmful activities such as visiting hazardous or unauthorized places by tracking worker movement.
4. Connected logistics: Optimize end-to-end logistics operations by detecting anomalies and deviations for private and third-party carriers, as well as warehousing facility operations.
5. Connected assets: Increase the value of acquisitions by improving maintenance efficiency. Optimize asset availability and utilization by measuring remotely and visualizing consumption, condition, performance, and operating environments.
6. Other applications of IoTs are like Smart Homes, Smart City, Self-driven Cars, IoT Retail Shops, Farming sector or Agriculture sector, Smart Grids, Smart Supply-chain Management, etc.

Intelligent Internet of Underwater Things (IoUT) Protocols and Standards for Intelligent IoT

The Internet of Underwater Things (IoUT) is a new type of IoT that is described as a network of intelligent underwater items that are all connected. Environmental monitoring, underwater research, and disaster avoidance are just a few of the practical uses that the Internet of Things is intended to enable. IoUT is recognized as one of the viable technologies for constructing smart cities because of these applications. Underwater Wireless Sensor Networks (UWSNs) have developed as a trusting network solution to assistance the idea of IoUT.

In the last decade, researchers were trying to present various practical and potential IoUT applications. Here, in the above Fig. 6, they classified the applications into five categories: (1) environmental system, (2) underwater process, (3) disaster prevention techniques, (4) military system, and (5) others.

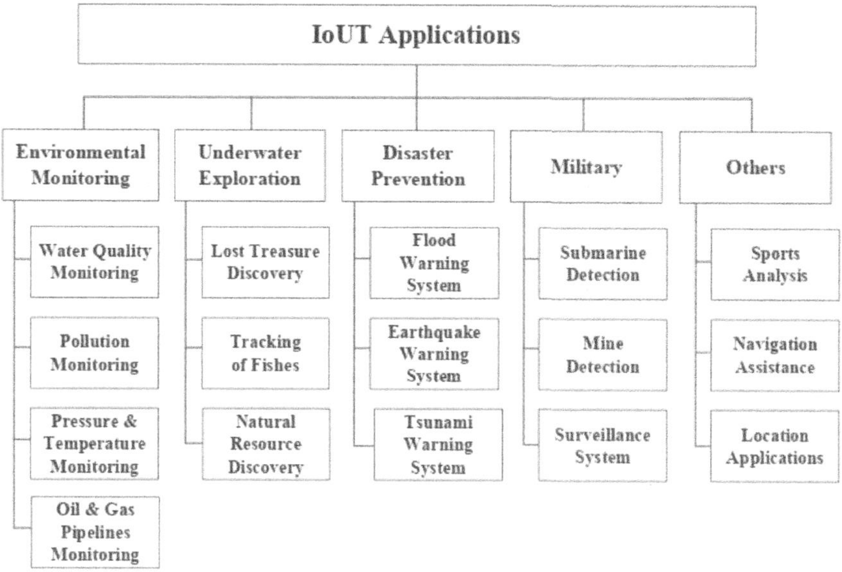

Fig. 6 The applications of IoUT [8]

Intelligent Internet of Vehicles

At present, the Internet technology system vehicles, i.e., Internet of Vehicle (IoV), have arisen from the IoT system, which connects vehicles in a given context to various parts of the society. Entirely autonomic driving vehicles and semi-autonomic driving vehicles make up the vehicles in the IoV ecosystem. In the IoV system, these vehicles communicate with both the external and interior environments. In the future, each vehicle is expected to be equipped with more than 200 sensors. Millions of vehicles will soon begin full operation in the IoV environment, generating over 25 GB of data per hour, resulting in so-called big data, which comes in various forms: structured form, semi-structured form, and unstructured form, and necessitates collecting data, processing data then manage it after that store and lastly analyze for decision-making. Because the IoV idea is yet in its initial phases, more thorough research is required to comprehend it fully. Few AI approaches have been employed to solve challenges and autonomous cars, leaving a gap that need to fill.

References

1. https://www.tutorialspoint.com/artificialintelligence/artificial_intelligent_systems.htm
2. https://whatis.techtarget.com/definition/intelligent-system
3. Ganesh Babu, R., Karthika, P., Aravinda Rajan, V.: Secure IoT Systems Using Raspberry PiMachine Learning Artificial Intelligence. ICCNCT 2019, LNDECT 44, pp. 797–805 (2020)
4. https://www.javatpoint.com/iot-architecture-models

5. https://jp.globalsign.com/repository/common/pdf/GlobalSign_CA_CPS_v9.5.pdf
6. Hameed, K., Sarwar Bajwa, I., Ramzan, S., Anwar, W., Khan, A.: An Intelligent IoT Based Healthcare System Using FuzzyNeural Networks. Hindawi Scientific Programming Volume 2020, Article ID 8836927, 15 pages
7. Ullah, I., Ahmad, S., Mehmood, F., Kim, D.: Cloud Based IoT Network Virtualization for Supporting Dynamic Connectivity among Connected Devices. Electronics **8**, 742 (2019). Doi:10.3390/electronics8070742. https://www.mdpi.com/journal/electronics
8. Kao, C., Lin, Y., Wu, G., Huang, C.: A Comprehensive Study on the Internet of Underwater Things: Applications, Challenges, and Channel Models. Sensors **17**, 1477 (2017). Doi:10.3390/s17071477. https://www.mdpi.com/journal/sensors

Contents

ML in WSN Using IoT for Smart Cities: A Survey

Piyush Raja

1 Introduction

A smart city is an urban region that collects data from many areas using remote sensors, and Internet of Things (IoT) enabled technology to improve people's quality of life. Wireless sensor networks with low power and low data rates are utilized in smart city monitoring and control applications. In the Internet of Things, WSN nodes serve as the underlying technological backbone. The "things" in the Internet of Things are small embedded physical sensing devices (i.e., WSN nodes) that are connected to the internet to execute a specified task. Artificial intelligence (AI) and machine learning (ML), two new breakthrough techniques, are now emerging as the future of fully automated IoT applications [1]. Machine learning is a branch of artificial intelligence in which computer systems learn on their own by improving on previous experiences. Until 2013, a comprehensive study of machine learning algorithms was conducted. Because ML and IoT technologies are quickly evolving, the authors have expanded their survey chapter. Smart traffic monitoring, smart grids, smart waste management, smart agriculture, smart medical healthcare, and other IoT applications in smart cities are examples. Table 1 lists all of the significant abbreviations used in this chapter in complete form.

Truly automated operation, maximum network lifetime, energy efficiency, quality of service (QoS), cross-layer optimization, high bandwidth demand, sensor data processing, cloud computing, communication protocol design, and other issues plague WSN-based IoT (WSN-IoT). The industrial Internet of Things (IIoT) or industry 4.0 is now the most significant transformation in smart industries, smart manufacturing, automobiles, smart cities, and medical healthcare. Various large corporations, such as Microsoft, Google, and Amazon, are developing AI and

P. Raja (✉)
Department of Computer Science, Magadh University, Bodh Gaya, Bihar, India
e-mail: piyushraja2009@gmail.com

S. Mukherjee et al. (eds.), *Intelligent Systems for Social Good*,
Advanced Technologies and Societal Change,
https://doi.org/10.1007/978-981-19-0770-8_1

Table 1 Shows an alphabetical list of abbreviations

Acronym	Description
5G	5th Generation Cellular Networks
6G	6th Generation Cellular Networks
AMQP	Advanced Message Queuing Protocol
ANN	Artificial Neural networks
BLE	Bluetooth Low Energy
CoAP	Constrained Application Protocol
DDS	Data Distribution Service
DL	Deep Learning
DT	Decision Tree
IoT	Internet of Things
k-NN	K Nearest Neighbourhood
LDA	Linear Discriminant Analysis
LoRaWAN	Long Ranged Wide Area Network
LTE	Long Term Evolution
MAC	Medium Access Control
MLMDP	Machine Learning Markov Decision Process
MLP	Multi-Layer Perceptron
MQTT	Message Query Telemetry Transport
NB-IoT	Narrowband IoT
PCA	Principle Component Analysis
QoS	Quality of Service
RL	Reinforcement Learning
RFID	Radio frequency Identification
SARSA	State-Action-Reward-State-Acton
SVM	Support Vector Machines
TCP	Transmission Control Protocol
UDP	User Datagram protocol
WSN-IoT	WSN based IoT
WSN	Wireless Sensor Networks

machine learning-based algorithms for sophisticated IoT applications in smart cities across the world.

Machine learning may be used in WSN-IoT for dynamic routing table updates, node localization in mobile WSN-IoT nodes, identifying and separating defective nodes for network optimization, and predicting the quantity of energy harvested in energy harvesting WSNs (EH-WSN). The writers of this study attempted to answer the following research questions: Why are machine learning techniques employed in the WSN-IoT? What are the advantages of utilizing machine learning over traditional optimization approaches in WSN-IoT? Why is the smart city a common IoT application use case?

Smart cities may now utilize data to better control traffic, decrease pollution, and make better use of infrastructure thanks to the Internet of Things [2]. The benefits of applying machine learning in conventional WSN-IoT are as follows:

In most cases, WSNs are deployed in a constantly changing environment. As a result, a fully automated IoT scenario should expect self-adaptation to the new environment. Unknown parameter monitoring, such as temperature measurement in a glacier or volcano monitoring, necessitates automatic network architecture and configuration adjustments. In WSN-IoT, there aren't any reliable mathematical models for the unknown parameters. Because WSN-IoT works with a lot of sensor data, the correlation between multiple data sets might be a big problem. WSN integration with IoT via cloud-based services for improved monitoring and control.

In the WSN-IoT, future predictions and actions are conceivable.

From millions of sensor nodes, the Internet of Things creates a tremendous amount of data. Data drive machine learning, which creates usable knowledge from prior data. Machine learning makes use of historical IoT data to uncover hidden patterns and creates models that may be used to forecast future behaviour and occurrences. Due to the resource constraints of WSN-IoT, there are several restrictions to executing ML-based inferences on IoT nodes, such as:

As more data are processed, a greater number of computations are required, increasing computation complexity.

Extra energy consumption.

Complex procedures and multi-domain expert programmers are required to train WSN-IoT nodes for multiple ML algorithms.

The contributions of this survey chapter in the realm of WSN-IoT are as follows:

The use of machine learning techniques as an optimal solution for classic WSN-IoT issues in smart cities is presented in this chapter. The WSN-IoT framework's design recommendations based on AI and ML have been presented [3]. For ML engineers and data scientists, an in-depth literature study of WSN-IoT in smart cities is provided in detail.

2 WSN-Based IoT

Figure 1 depicts the functioning of WSN-IoT in a smart city. The WSN nodes are used in smart city applications including traffic monitoring, creating smart grids, remote health care monitoring, smart agriculture, and industrial applications. A smart city's IoT-WSN node's job is to continually monitor and manage physical quantities including temperature, humidity, pressure, acceleration, and so on. These sensor nodes' main job is to collect data and deliver it to the main WSN-IoT gateway node. The data are transmitted to the cloud server from the gateway node [4]. Cloud computing takes occur in the IoT cloud. Remote servers, user mobile

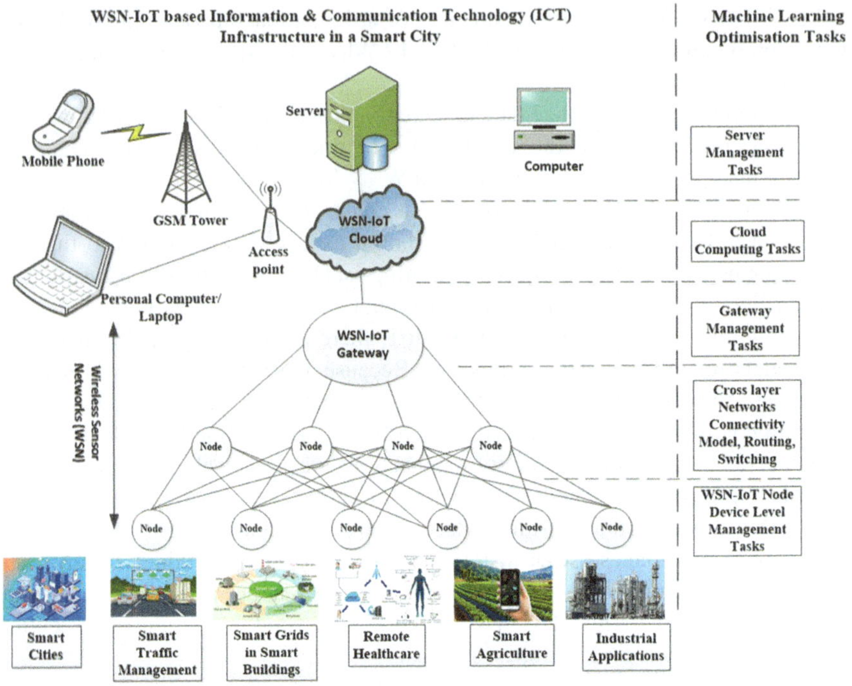

Fig. 1 WSN-IoT in a smart city utilizing ML

phones, computers, mobile phone towers, and other IoT cloud components are all directly connected. IoT and machine learning operations need a significant quantity of data processing and memory. As a result, the IoT cloud server is built as a high-performance, high-processing computer with a large storage capacity. WSN end nodes, on the other hand, have limited computational capabilities, limited storage, and a finite non-rechargeable battery power source. The maximum data rate for WSN-IoT end nodes based on the IEEE 802.15.4 standard is 250 kbps. End nodes of WSN-IoT are powered by two AA-size batteries (1.5 V, 1000 mAh), while the gateway is supplied by the mains. Machine learning algorithms may be deployed from the cloud to the WSN-IoT node device level for autonomous operation. The sensor data are sent to the cloud server via the IoT device [5]. The user may monitor and manage the application from the IoT cloud using a smartphone, laptop, or desktop PC, as well as a personal digital assistant. Many prominent cloud service providers now provide a free, but restricted, quantity of sensor data storage in their cloud storage. For example, Mathworks Incorporation's Microsoft Azure IoT, Amazon Web Service (AWS), Google cloud platform, Cisco IoT cloud connect, IBM Watson IoT, and Thing talk IoT.

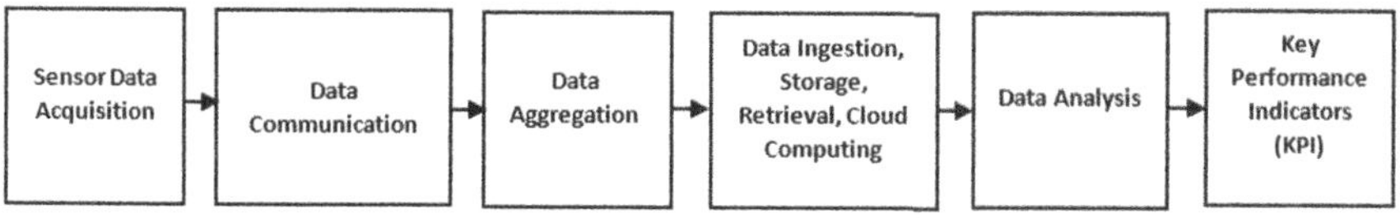

Fig. 2 A typical WSN-IoT application's data flow

3 Preliminaries in ML for WSN-IoT

ML is a branch of artificial intelligence that allows computers to learn and develop on their own without having to be explicitly programmed. Its goal is to create new computer programs that can access data and learn on their own (Figs. 2 and 3).

I. **Training Process**: First and foremost, data are obtained through a certain application. These raw data are used to extract the characteristics. If the data includes picture data, for example, the colours, pixels, brightness, and contrast of the whole database of images are extracted. The features are then categorized according to the machine learning process's requirements [6]. In order to learn or enhance the fundamental starting algorithms, certain training examples are now applied to them. As a result, algorithms are taught and optimized based on data patterns.

II. **Testing Process**: The next stage is to put this well-trained WSN to work in a real-world application. In the actual world, unknown data are used as input and characteristics are retrieved [7]. These retrieved characteristics are fed into an algorithm that has previously been trained. The trained algorithm's output is categorized as data predictions.

III. **Actions of the WSN**: Finally, the WSN determines the necessary actions based on expected output data.

4 Open Research Problems in WSN-IoT That ML Techniques Can Solve

The following are open research topics in WSN-IoT that can be solved using machine learning approaches.

5 Localization of IoT Nodes

The current position identification of a sensor node in a WSN situation is referred to as node localization. Path planning is a critical stage in mobile WSN nodes. Because all nodes are split (classified) into range-based and range-free nodes, node localization is considered a classification challenge. For node localization as a classification issue,

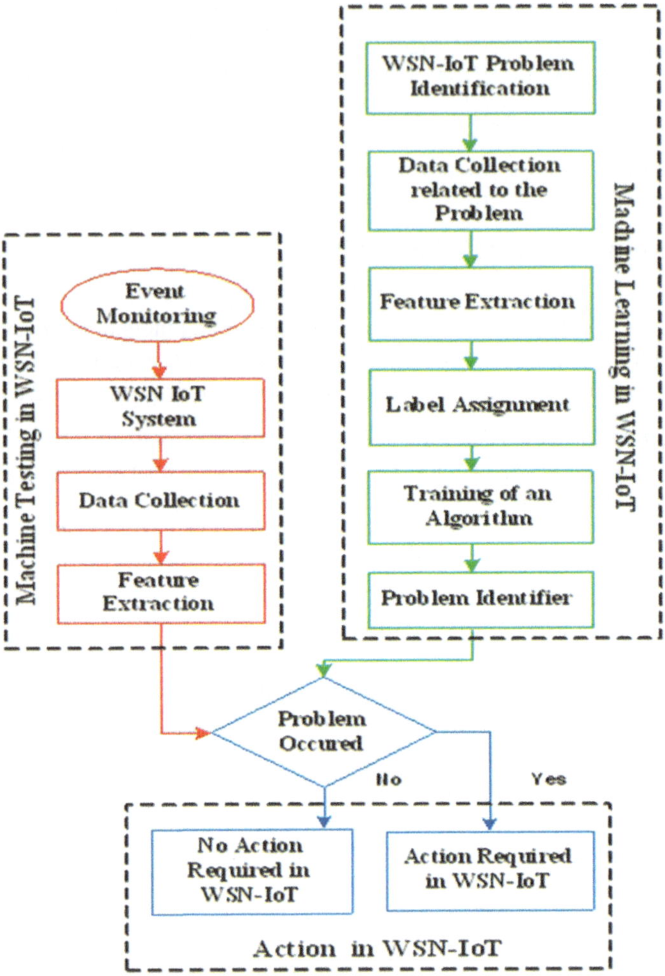

Fig. 3 Machine learning process flowchart in WSN-IoT

several ML algorithms such as SVM, K-NN, and RL-based approaches (Q-learning, SARSA) are employed in WSN-IoT.

6 Coverage and Connectivity of IoT Nodes

The sensing coverage in a WSN scenario is the field of interest (FOI), which is defined as the area covered by at least one sensor node. As a result, the best location for sensor nodes is a design challenge [8]. The connection between neighbour nodes should be suitable to enhance the WSN lifespan.

7 Issues of Routing Layer

Routing is the process of delivering data packets from one node to another via intermediary nodes [9]. Long routing tables, which contain the source and destination addresses of all packets in the network, are maintained by gateway nodes during the routing process [10]. The end nodes of a WSN send the detected data to the main gateway node. In a WSN network, if the routing path is excessively lengthy, a lot of energy is lost. As a result, smart routing algorithms must be carefully developed to identify the best paths between end nodes and gateway nodes [11]. WSN uses a variety of machine learning approaches to identify the best path, including decision trees, random forests, ANN, SVM, and Bayesian learning [12].

8 Issues with the MAC Layer

WSN's media accessing method is controlled by the MAC layer. In WSN, the sensor MAC (SMAC) protocol is commonly employed [13]. In WSN, methods based on reinforcement learning (RL) are utilized to build MAC protocols. In sensor networks, RL-MAC methods manage sleep, waking, transmission, and reception [14].

9 Aggregation of Sensor Data

Thousands of tiny sensor nodes are distributed in a smart city to measure the same physical quantity, such as temperature, humidity, light, CO_2 gases, and so on. Several sensors might send the same data to the gateway. The gateway struggles to process such a big volume of data. As a result, in WSNs for smart city applications, sensor data aggregation is critical. Data aggregation is the process of gathering and combining valuable data from many sources [15]. Data redundancy and accuracy are increased as a result of this procedure. Data aggregation reduces the power consumption of WSN nodes, extending the network's lifespan. Data aggregation can benefit from machine learning [16]. The data from the cluster head are aggregated and sent to the base station by the cluster [17]. For data aggregation tasks in WSN-IoT, machine learning approaches based on artificial neural networks (ANN) and quality (Q)-learning algorithms are effective [18].

10 Target Detection and Event Monitoring

WSNs are used in smart city applications to monitor events and detect targets, such as intrusion detection and traffic tracking [19]. Node failure, target recovery, and sensing node tracking latency are all necessary for a WSN. For event monitoring and target tracking in WSNs, several ML approaches such as Bayesian, Q learning, and genetic algorithms are utilized [20]. Using machine learning techniques in WSNs can help recognize an event or target from complicated image sensor data.

11 Harvesting of Energy

The process of gathering environmental energy from the sun, wind, tides, radio waves, and other sources and converting it into electrical energy is known as energy harvesting [21]. The overall goal of energy harvesting is to conserve our finite supply of fossil fuels (coal, oil and gases). Energy harvesting, on the other hand, may be utilized in smart city applications to achieve maximum network lifespan in rechargeable battery-based WSN-IoT nodes [22]. Furthermore, machine learning techniques are employed in energy harvesting WSN-IoT activities for predicting future available energy [23]. For energy harvesting applications, machine learning methods such as regression and reinforcement learning approaches (Q-learning) are applicable. With rechargeable battery-based WSNs, solar energy, radio frequency waves, and wind energy are typically employed. Traditional WSN-IoT activities such as harvested energy forecast and battery power management can benefit from the use of machine learning techniques.

12 Processing of Node Query

End nodes, cluster heads, and gateway nodes all conduct different sorts of queries in WSNs, including sensor data aggregation, routing pathways, synchronization and control activities, packet delivery, and so on [24]. For sensor data queries in WSN, k-nearest neighbourhood-based ML methods are employed.

13 Our ML Techniques in WSN-IoT Survey Report

We conducted this chapter by looking up machine learning in WSN-IoT for smart cities on numerous websites, journals, magazines, and research papers [25]. We found several articles accessible from 2010 to 2021 by searching the internet for ML methods tackling WSN-IoT issues [26]. It illustrates the percentage contribution of category-wise ML algorithms solving WSN-IoT problems. The graphical depiction of main ML algorithms in WSN-IoT is shown in Fig. 4.

Figure 5 depicts a brief overview and simple graphical depiction of the key machine learning techniques utilized in WSN-IoT.

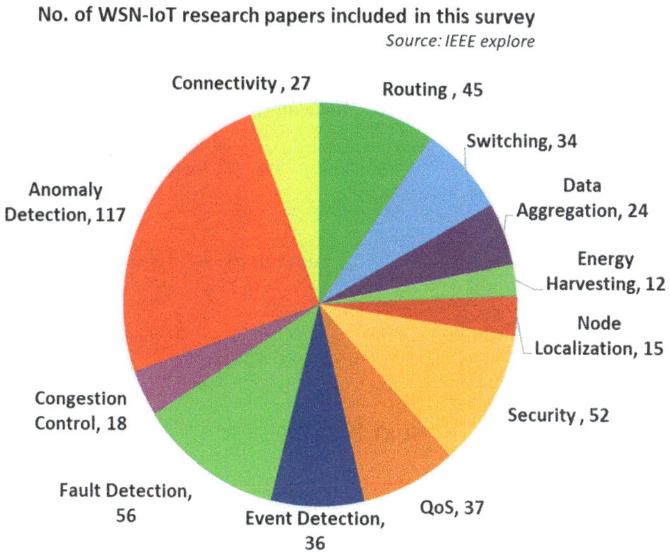

Fig. 4 Shows the research articles on WSN-IoT technologies. *Source* IEEE explore

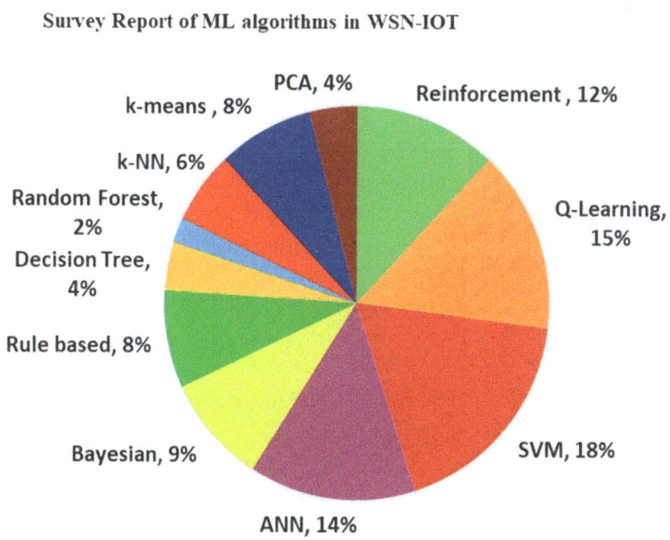

Fig. 5 Graphical representation of major ML algorithms used in WSN-IoT. *Source* IEEE explore

Figure 6 depicts the prevalence of RL, supervised, and unsupervised learning algorithms in WSN-IoT in a simple graphical depiction.

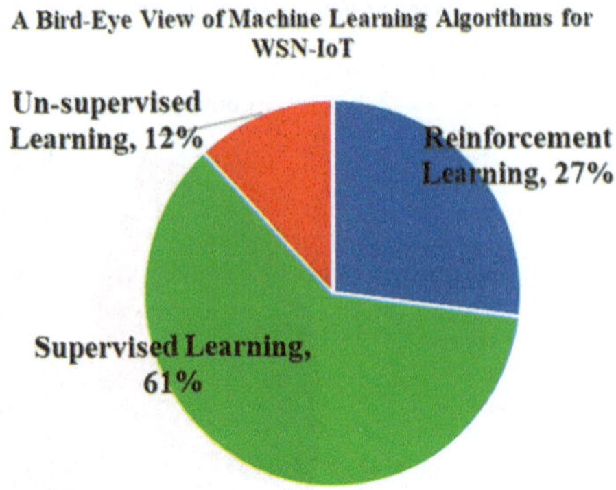

Fig. 6 Graphical representation of RL, supervised and unsupervised learning algorithms in WSN-IoT. *Source* IEEE explore

14 Conclusions and the Next Steps

We discussed several machine learning methods in WSN-IoT for smart city applications in this chapter. We conducted a thorough assessment of ML methods in WSN-IoT for smart city issues in the chapter. According to the results of this poll, supervised learning algorithms were employed the most, with 61%, compared to RL with 27% and unsupervised with just 12%. Because machine learning algorithms are so diverse and strong, they can be utilized for many tasks in WSN-IoT in smart cities. In WSN-IoT, the sophisticated SVM algorithm, for example, may be utilized for classification and regression applications. A more powerful and complicated algorithm will develop in the future, reducing the need for human involvement. Machine learning techniques will be used in the upcoming IoT-based smart city solution. The heart stroke rehabilitation system in smart health care, for example, employs LDA, MLP, and SVM algorithms. Ultra-dense cellular IoT networks based on high-performance machine learning algorithms will be used in next-generation smart cities.

References

1. Alsheikh, M.A., Lin, S., Niyato, D., Tan, H.-P.: Machine learning in wireless sensor networks: algorithms, strategies, and applications. IEEE Commun. Surv. Tutor. **16**, 1996–2018 (2014)
2. Du, R., Santi, P., Xiao, M., Vasilakos, A.V., Fischione, C.: The sensable city: a survey on the deployment and management for smart city monitoring. IEEE Commun. Surv. Tutor. **21**, 1533–1560 (2019)
3. Morello, R., Mukhopadhyay, S.C., Liu, Z., Slomovitz, D., Samantaray, S.R.: Advances on sensing technologies for smart cities and power grids: a review. IEEE Sens. J. **17**, 7596–7610 (2017)
4. Sharma, H., Haque, A., Jaffery, Z.A.: Modelling and optimization of a solar energy harvesting system for wireless sensor network nodes. J. Sensor Actuator Netw. **7**, 1–19 (2018)
5. Mackey, A., Spachos, P., Plataniotis, K.N.: Smart parking system based on bluetooth low energy beacons with particle filtering. IEEE Syst. J. **14**, 1–12 (2020)

6. Nikoukar, A., Raza, S., Poole, A., Güneş, M., Dezfouli, B.: Low-power wireless for the internet of things: Standards and applications. IEEE Access **6**, 67893–67926 (2018)
7. Yan, Y., Sharif, A., Ouyang, J., Zhang, C., Ma, X.: UHF RFID handset antenna design with slant polarization for IoT and future 5G enabled smart cities applications using CM analysis. IEEE Access **8**, 22792–22801 (2020)
8. IEEE Standard for RFID. Local and metropolitan area networks IEEE 802.15.4: Low-rate wireless personal area networks (LR- WPANs). In: Amendment 2: Active Radio Frequency Identification (RFID) System Physical Layer (PHY), IEEE Std 802.15.4; IEEE: Piscataway, NJ, USA, 2012
9. Anagnostopoulos, T., Zaslavsky, A., Kolomvatsos, K.: Challenges and opportunities of waste management in IoT-enabled smart cities: a survey. IEEE Trans. Sustain. Comput. **2**, 275–289 (2017)
10. Iglesias-Urkia, M., Casado-Mansilla, D.: Integrating electrical substations within the IoT using IEC 61850, CoAP, and CBOR. IEEE Internet Things J. **6**, 7437–7449 (2019). [CrossRef] Technologies. IEEE Internet Things J. **5**, 784–793 (2018)
11. IEEE Standard for ZigBee. Low-Rate Wireless Networks—Amendment 3: Use of the 865 MHz to 867 MHz Band in India, Amendment to IEEE Std 802.15.4; IEEE: Piscataway, NJ, USA, 2017
12. Luzuriaga, J.E., Perez, M.: A comparative evaluation of AMQP and MQTT protocols over unstable and mobile networks. In: Proceedings of the 2015 12th Annual IEEE Consumer Communications and Networking Conference (CCNC), Las Vegas, NV, USA, 9–12 January 2015
13. Lachtar, A., Val, T., Kachouri, A.: Elderly monitoring system in a smart city environment using LoRa and MQTT. IET Wirel. Sens. Syst. **10**, 70–77 (2020)
14. Zhang, D., Chen, Z., Ren, J., Zhang, N., Awad, M.K., Zhou, H., Shen, X.S.: Energy-harvesting-aided spectrum sensing and data transmission in heterogeneous cognitive radio sensor network. IEEE Trans. Veh. Technol. **66**, 831–843 (2017)
15. Li, H., Dong, W.: Enhancing the performance of 802.15.4-based wireless sensor networks with NB-IoT. IEEE Internet Things J. **7**, 3523–3534 (2020)
16. Sharma, H., Haque, A., Jaffery, Z.A.: Solar energy harvesting wireless sensor network nodes: a survey. J. Renew. Sustain. Energy **10**, 1–33 (2018)
17. Rostami, S., Heiska, K., Puchko, O., Leppanen, K., Valkama, M.: Pre-grant signaling for energy-efficient 5G and beyond mobile devices: Method and analysis. IEEE Trans. Green Commun. Netw. **3**, 418–432 (2019)
18. Chettri, L., Bera, R.: A comprehensive survey on internet of things (IoT) toward 5G wireless systems. IEEE Internet Things J. **7**, 16–32 (2020)
19. Davoli, L., Belli, L.: From micro to macro IoT: Challenges and solutions in the integration of IEEE 802.15.4/802.11 and Sub-GHz te
20. Kolias, C., Kambourakis, G., Gritzalis, S.: Attacks and Countermeasures on 802.16: Analysis and assessment. IEEE Commun. Surv. Tutor. **15**, 487–514 (2013)
21. Horng, G.-J., Liu, M.-X., Chen, C.-C.: The smart image recognition mechanism for crop harvesting system in intelligent agriculture. IEEE Sens. J. **20**, 2766–2781 (2020)
22. Pirayesh, H., Kheirkhah, P.: Coexistence of Wi-Fi and IoT communications in WLANs. IEEE Internet Things J. **7**, 7495–7505 (2020)
23. Pal, D., Funilkul, S., Charoenkitkarn, N., Kanthamanon, P.: Internet-of-things and smart homes for elderly healthcare: an end user perspective. IEEE Access **6**, 10483–10496 (2018)
24. Sundaram, J.P.S., Du, W., Zhao, Z.: A Survey on LoRa networking: research problems, current solutions, and open issues. IEEE Commun. Surv. Tutor. **22**, 371–388 (2020)
25. Chang, T.-C., Lin, C.-H.: Traffic-aware sensor grouping for IEEE 802.11ah networks: Regression-based analysis and design. IEEE Trans. Mob. Comput. **18**, 674–687 (2019)
26. Klymash, M., Beshley, H.: Improving architecture of LTE mobile network for IoT services provisioning. In: Proceedings of the 2nd IEEE International Conference on Advanced Information and Communication Technologies (AICT), Lviv, Ukraine, 4–7 July 2017

Prospects of 'SMART Farming' in Cold Arid Region of Ladakh, India

S. Angchuk and Aleem Ali

1 Introduction

The Union Territory of Ladakh with its distinct geography and climatic conditions altogether offer a tough life managed with meager resources. The agriculture sector being the primary occupation of this region has, directly and indirectly, contributed towards the sustenance of the region. The new era of change and development also demands the advancement in the farming sector of the region with advanced technology and inputs. While working in the crop fields, farmers are required to carry out several responsibilities. Seeding, weeding, fertilizing, and watering, for example, appear to be repetitive and labor-intensive operations undertaken by farmers in the field. However, to make the farming cycle more effective, preliminary decision-making is required before commencing the actual activities. This is being taken care of under a set of interrelated technologies that take care of the farming needs right from seed selection to sowing, harvesting, storing, processing and value addition, marketing, weather parameters, and all other relevant needs of the farming communities. This set of interrelated technologies also aids in decision-making and is referred to as Smart Farming. Smart farming is a type of agriculture management that makes use of contemporary technologies to boost the quantity and quality of agricultural products [1]. Smart agriculture helps to address the aforementioned problems by reducing crop waste, making better use of fertilizer thereby increasing the yield of the crop. Smart Farming is being practiced using different tools and technologies.

The Internet of Things (IoT) is a network of interconnected computing devices, digital and mechanical equipment, humans or animals, and objects that can perceive,

S. Angchuk (✉) · A. Ali
Department of Computer Science, Glocal University, Saharanpur, UP, India
e-mail: angchuks@gmail.com

A. Ali
e-mail: aleem@theglocaluniversity.in

S. Mukherjee et al. (eds.), *Intelligent Systems for Social Good*,
Advanced Technologies and Societal Change,
https://doi.org/10.1007/978-981-19-0770-8_2

gather, and transport data over the internet without the need for human intervention. It is a technology that aims to link all of the world's items to the Internet. The Internet of Things' main objective is to link everything in our environment and enable seamless communication between them with minimal human involvement. It emphasizes the ability to connect with anybody, anywhere, at any time. It allows real-life physical objects (e.g. sensor nodes) to work together to develop an information-based system that maximizes the benefits (e.g. improved agricultural production) with minimized risks (e.g. environmental impact).

IoT helps in predicting crop production, crop pricing, the temperature of the soil, and real-time data on air quality, water level, and crop delivery timing, all of which helps in boosting the productivity of the crop. As the world's population grows, so does the demand for food, and IoT in agriculture is a key driver in meeting that demand. While IoT-based agriculture is thriving in developed countries, in India, it is still in its infancy. Lack of awareness about technical equipment among the farmers is the major challenge that we are facing. Furthermore, in India, the cost of implementation is also a big issue [2]. As a result, we should concentrate on building more specialized and effective sensors and implementing them using the right technique (Fig. 1).

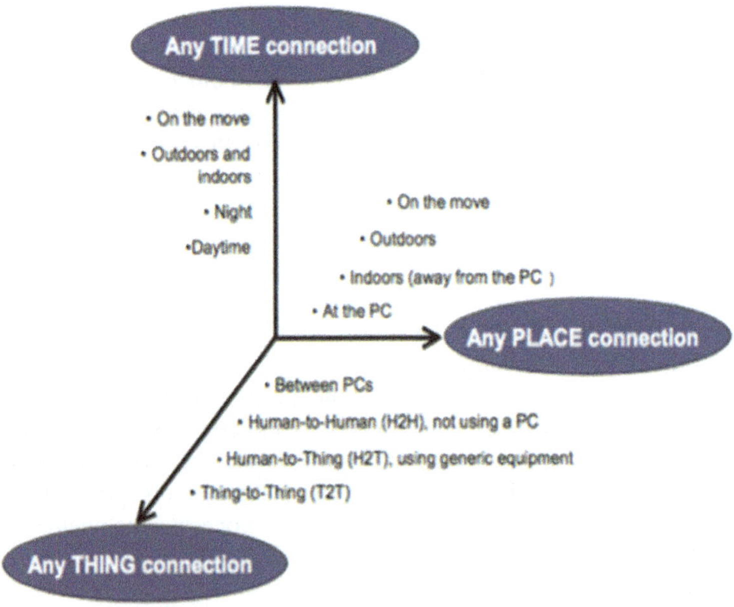

Fig. 1 Internet of Things [2]

2 The Problem Faced by Farmers

Like in the rest of the state, agriculture is the main source of livelihood in this district. Of the 45,167 ha, the area under cultivation in this district is 10,542 ha as per the revenue department village paper. As land holdings are small (0.68 ha) [3], more people are working invariably on farms in rural areas, and with conventional manual methods, outdated equipment limits the production thereby reducing revenues from farming. The main problems faced by the farmers of this region are mentioned below [4].

- Very short growing season, extreme climate, harsh winters, high-temperature variation, extremely arid conditions, negligible rainfall, and inhospitable weather conditions.
- Geographically difficult zone, demographically disadvantaged, and politically sensitive.
- Scarcity of high-quality planting material both in agricultural as well as horticultural crops.
- Inadequate irrigation and electricity supply.
- The fertility status of the soil is low.
- Inadequate technical know-how among farmers for improvement in agriculture and allied sector.
- Poor marketing facility (due to which local produce does not fetch a genuine price.).
- Lack of network, cooperative societies for storage, marketing, supply of farm inputs, etc.
- Lack of feed and fodder.
- Lack of mechanization in agriculture.
- Inaccessibility to villages hampers in transferring of technology.
- The district remains cut off from the rest of the country for more than 6 months.

3 Smart Agriculture

The modern practice of advanced technology and equipping the farm is conclusively termed Smart Farming or Smart Agriculture. Smart farming is a modern information and communication technology idea that aims to enhance the quantity and quality of goods to control farm management. IoT-controlled smart farming can be called a wireless system designed to automate the irrigation process and to monitor the field using different sensors. These different sensors may be light, temperature, humidity, soil, etc. [5]. Smart farming (or precision agriculture) has the potential to significantly increase agricultural production in terms of both sustainability and productivity [6]. The adoption of smart farming techniques is constantly growing, and the market is still dynamic for these connected devices. By adopting various smart agriculture gadgets, farmers have become more efficient in raising livestock and growing crops,

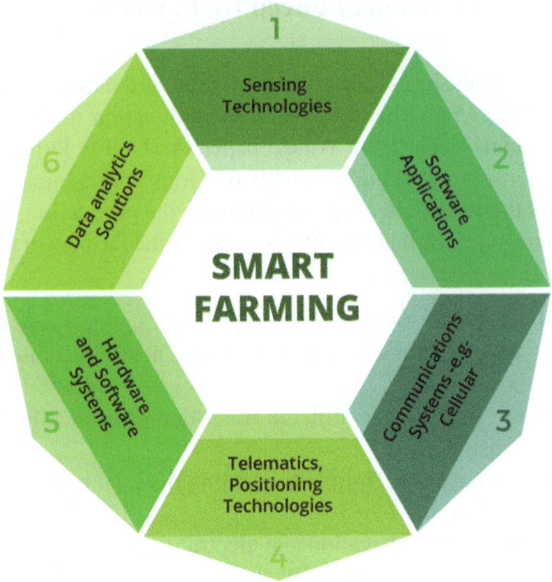

Fig. 2 Smart farming [7]

making it more predictable and efficient. Smart farming is a term that refers to the use of modern technologies in agriculture. The smart agriculture industry is growing very fastly and constantly. By using the advanced technology, they are gaining better control over the complete production efficiently [7] (Fig. 2).

Smart technologies are not only confined to the agriculture sector but these find a wide application in other sectors too. These technologies are used to locate the location of the accident and the information of the location can be sent through the GPS to the emergency offerings for assistance [8]. Data mining is another field where the use of smart technology is on the rise. Data mining is an analytic process that is used to investigate data (business or market related) in search of consistent patterns or to find out the systematic relationship between variables and to validate the findings [9].

Greenhouses play an important role in agriculture and gardening as they allow plants to grow under regulated climatic conditions. The disadvantage of the hand-operated greenhouse is that it always requires inspection of plant growth visually, turning on and off the temperature controllers manually. The environmental factors within a greenhouse may be monitored and controlled with the help of an IoT-based greenhouse. It offers a controlled environment for the plants, preventing damage and so improving total yield. By maintaining any climatic condition in these greenhouses, it is possible to cultivate any type of crop [10].

Among the advanced smart farming technologies, some available technologies in the present scenario are:

- Sensor-based technology: soil scanner, water, and light management, temperature control, etc.

- Software-based technology: specialized software solution system.
- Communication-based technology—cellular communication.
- Positioning-related technology: including Global Positioning System.
- Hardware and software-related systems: it is related to robotics and automation.
- Data analysis-related technology: decision-making and prediction processes.

4 Major Factors for Smart Farming Using IoT

IoT has transformed the old agricultural perspective into advanced agriculture, which is geared towards the information network and involves automation, the usage of intelligent devices, and their networking in the agricultural production process. The following are some of the key points for Smart farming

- An Irrigation System can be used to provide water to the plant as and when required. The moisture sensor detects moisture in the plant and supplies water appropriately, reducing water usage to a bare minimum. Smart sensors are implanted in the soil. This sensor can quickly detect moisture levels and helps the farmer in sprinkling the correct quantity of water on the land.
- Farmers can use mobile applications to remotely track and manage yields, costs, and other vital farm metrics.
- Sensing technology (on-field sensors, such as soil moisture measurement) has shown to be quite beneficial, and smart positioning technologies (GPS) for making agricultural methods smarter have grown in popularity.
- When combined with weather data, it may provide farmers with weather forecasts.
- Set temperature, humidity, and other variables for agricultural storage to trigger alerts and alarms.
- Telematics (the transmission of data across vast distances) and advanced data analytics tools and platforms have also been important components of smart agriculture [11].

5 Uses of Internet of Things (IoT) in Agriculture

Farmers that rely on manual methods for crop monitoring, disease detection, and other tasks have several disadvantages, including the fact that it takes a long time, that they must physically show themselves on the farm, and that they are unable to identify the precise condition [12]. Farmers want very fundamental information of agriculture such as soil information, seed type, required pesticide for the specific crop at all stages of growth, fertilizer type, crop diseases, and crop selling. The following are the questions that must be addressed to improve crop output.

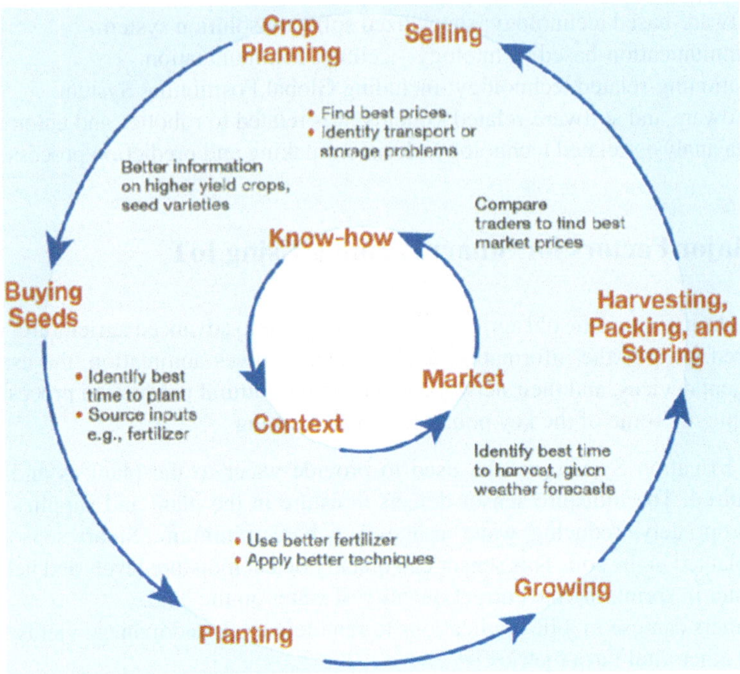

Fig. 3 IoT application in agriculture [14]

- Basic knowledge, such as which crop to plant?
- What seed varieties should be used?
- What weather information is needed?
- Best farming practices for his crops and soil.
- What kind of fertilizer and insecticides will be necessary for the crop?
- Transport prices, demand indicators, and logistical information [13] (Fig. 3).

Smart farming is a high-tech and efficient method for sustainable farming and food production in agriculture. It is a technique of integrating connected devices and innovative technologies into agriculture. Smart farming is highly reliant on IoT, which helps in eliminating the need for physical labor on the part of farmers and producers while improving production in every possible manner. With modern agricultural developments reliant on agriculture, the Internet of Things has provided significant benefits such as efficient water usage, input optimization, and many more. As the food demands of a growing global population put increasing strain on agriculture throughout the world, using IoT technology can drive farming efficiency to entirely new heights [15]. IoT-based smart farming strengthens the whole agriculture system by monitoring the land in real time. With the help of sensors and other gadgets, the Internet of Things in agriculture has not only saved farmers' time but also help in reducing the resources such as water and electricity. It keeps the track of a variety of factors such as humidity, temperature, and soil, among others, and provides an accurate result.

6 Sensors Used in Agricultural Internet of Things

Sensors are important tools used in IoT. Sensors are devices that gather and analyze data to produce the necessary analysis. Sensors are mostly used in agriculture to obtain readings for measuring NPK levels, detecting diseases, and determining soil moisture content.

(a) Soil Moisture Sensor

A soil moisture sensor is shown in Fig. 4. It is used to determine the volumetric water content of the soil. It determines the moisture content of the soil based on soil characteristics such as resistance, dielectric constant, and neutron interaction, as well as ambient factors such as soil type, temperature, and electrical conductivity. It has two probes that are placed into the field, and when a current is passed through the probes, moisture % is determined based on resistivity. Soil moisture analysis enables water to be applied just when it is required, reducing water waste.

(b) Temperature sensor

A temperature sensor is a device that measures how hot or cold an object is. This sensor is more accurate than the thermistor, which was used to monitor the temperature in the beginning. This sensor is resistant to overheating since it has three terminals: input, output, and ground. Temperature sensors come in a variety of shapes and sizes. The LM-35 IC, as illustrated in Fig. 5, is one form of a temperature sensor.

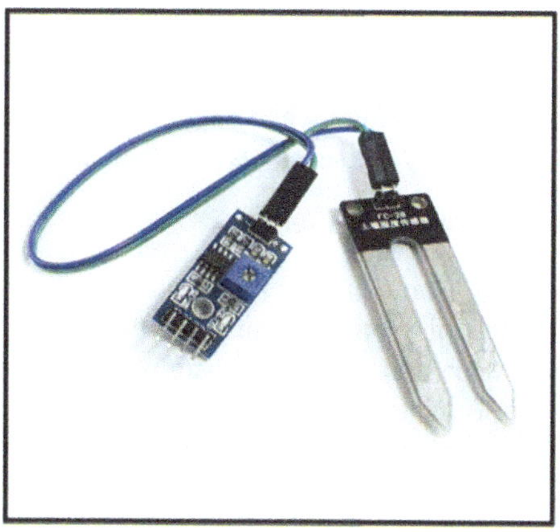

Fig. 4 Soil moisture sensor [16]

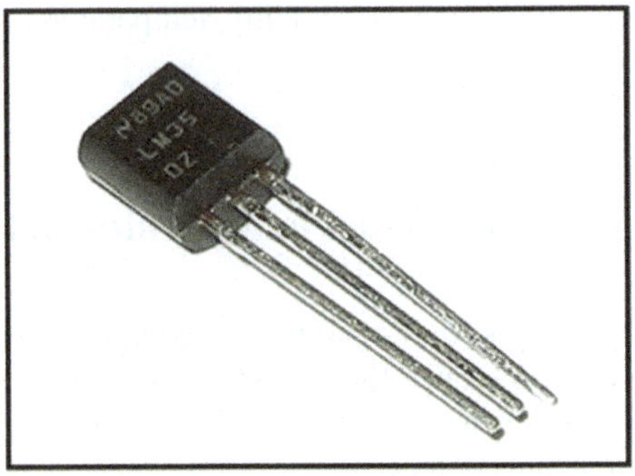

Fig. 5 Temperature sensor [16]

Fig. 6 PIR sensor [16]

(c) **Private Infrared (PIR) sensor**

All things with a temperature greater than absolute zero produce heat energy in the form of radiation. Figure 6 shows a PIR (private infrared) sensor that detects infrared radiation produced or reflected from an object. It is used to track the movement of humans, animals, and anything else. The temperature at the spot will rise from room temperature when any impediment passes through the field. It is converted to a voltage by the sensor, which then activates the detection.

(d) **Water level sensor**

Figure 7 shows a sensor that detects the amount of water or other fluids. It comes with a detecting probe that can detect the surface level of almost any fluid, including water, saltwater, and oils. This sensor is not readily broken, and it connects to Aurdino with ease. It includes two buttons, one for recording the lowest fluid level and the other for recording the highest fluid level. The voltage will be used to determine the level.

(e) **pH sensor**

The pH sensor in Fig. 8 is being used to measure the pH value of a solution. The pH scale ranges from 0 to 14, with 0–6 being acidic, 7 being neutral, and 8–14 being

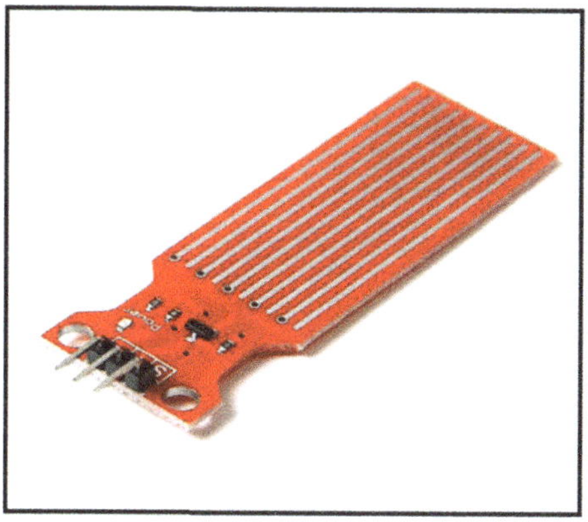

Fig. 7 Water level sensor [16]

Fig. 8 pH sensor [16]

non-acidic or basic. It calculates the pH value depending on the concentration of hydrogen ions, which is detected using a pH electrode. The response time is more than 2 min. The temperature range is about 600 °C, the input voltage is 5 V, and the output voltage is 414.12 V.

(f) **Temperature and humidity sensor**

In Fig. 9, the DHT11 is a simple, low-cost digital temperature and humidity sensor. This sensor has two components, a capacitive humidity sensor, and a thermistor. Both moisture and air temperature are sensed, measured, and reported by the humidity sensor. Temperatures range from 0 to 50 °C, with humidity levels ranging from 20 to 90%. These sensors are mostly used in the Internet of Things. There are many other sensors in addition to these; however, DHT11 is the most common temperature sensor.

Fig. 9 Temperature and humidity sensor [16]

7 Benefits of Adopting IoT in Agriculture [3]

(a) **Climate Condition**

In farming, climate plays a very important role. Climate change has a direct impact on all agricultural aspects. It has an immediate impact on crop quality and yield. As a result, an immediate solution to this problem is required. IoT solutions allow us to know the weather conditions in real-time. Sensors are placed both within and outside of agricultural areas. Environmental data are gathered by using these sensors, which are then used to choose the best crops for growing and sustaining under the given climatic conditions. The whole Internet of Things comprises sensors that can reliably monitor real-time weather variables such as humidity, rainfall, temperature, and more. A wide range of sensors is available to detect all of these characteristics and adjust them to meet the needs of smart farming. These sensors help in monitoring the health of crops as well as the weather conditions around them. When unexpected weather conditions are found, an alarm is sent. The necessity for human presence during inclement weather is removed, thus allowing farmers to enjoy additional agricultural advantages and enhance productivity (Fig. 10).

(b) **Disease Detection and Diagnosis**

Plants like any other organisms are vulnerable to many diseases. Diseases in plants occur when external factors like pathogens and environmental conditions disturb the normal natural structure and function of the plant. There is numerous type of diseases that affect plant species, and the factors responsible for these diseases are all different. Diseases may affect any part of the plant including root, stem, leaf, flower, and fruit. While disease grows within the plant, external factors like rain, wind, water, soil aid in transmitting these diseases to the neighboring plant. Many crops get spoiled due to a lack of proper pesticide control mechanisms. With the help of IoT enable systems, images of plant leaves are captured and investigated for

Fig. 10 Role of IoT in agriculture [3]

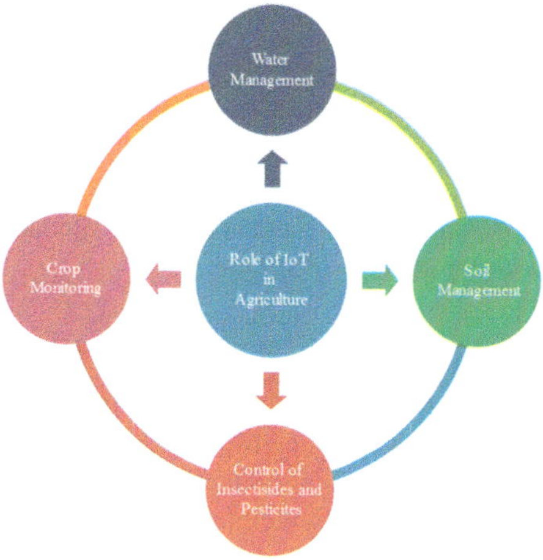

diseases, which can then be preprocessed and transmitted to remote laboratories. To decrease the cost of transmitting diseased leaf images to plant pathologists in remote laboratories, image preprocessing is required. Further clustering techniques aid in the segmentation of leaf images.

(c) **Fertilizer Calculator**

The soil fertility and soil type play very important roles in agriculture. Based on the type of soil, fertilizer can be injected into the soil for better crop yield production. Fertigation assures equal distribution of fertilizers to all of the vegetation and better absorption of nutrients. Based on recent technology, the climatic conditions like temperature, humidity, moisture, and Ph in the soil can be analyzed, and fertilizers like nitrogen, phosphorous, and potassium are injected into the flora. Fertilizer application is a crucial farming activity that can greatly boost agricultural production. For each crop, farmers must make judgments about which chemicals to use and in what quantities [17]. An IoT-based automated fertigation machine that reduces water and fertilizer wastage can be used to overcome these shortcomings.

(d) **Study of Soil**

Another essential aspect of farming that has a big influence on agricultural performance is soil. Soil monitoring with IoT allows the farmers and producers to enhance output, decrease disease, and optimize resources by leveraging technology. IoT sensors can be used to detect soil temperature, NPK, volumetric water, photosynthetic radiation, water potential of the soil, and soil oxygen levels. The data from the IoT sensors are subsequently sent to a centralized location (or the cloud) for analysis, visualization, and trend analysis. The information gathered may then be utilized to improve agricultural operations and make minor modifications to improve crop output and quality.

(e) **Crop water estimation and water study**

Water quality has a great impact on farming and agricultural output. Farmers must make prior decisions about how much water their crops require. Crop water needs are determined by several factors including crop type, season, climate, and crop growth stage. Crops lose water due to transpiration, while the canopy loses water due to evaporation. Using sensors, water management can be done efficiently using IoT with no wastage of water.

(f) **Analysis of Crop Produce Readiness**

Farmers that are provided with crop pricing information ahead of time may sell their harvests at a specified time and earn a high profit. To detect the freshness of fruits, a unique application of smartphone-based sensors is employed. To evaluate maturity levels and ripeness of green fruits, an IoT-based application, and a smartphone camera are used to collect pictures of fruits under white and UV-A light sources. Farmers should use these methods in their fields by sorting crops based on ripeness levels before shipping them to market.

8 Use of Smartphone in Agriculture

Information and communication technologies have played a significant influence in farmers' daily lives. ICT (Information and Communication Technology) in agriculture is a new discipline that focuses on agricultural and rural development. The use of ICT in agriculture allows for the timely distribution of critical information.

The fast expansion of mobile telephony and the advent of mobile-enabled information services have made it possible to overcome current information asymmetry in the fields of agriculture, healthcare, and education. There is a significant gap between the availability and distribution of agricultural inputs and agricultural infrastructure, which mobile technology can help to bridge. A smartphone is a device that is used to make phone calls and includes additional features and abilities such as the ability to send and receive an email, Wi-Fi and modem capabilities, internet access, office documents, easy touch screen control, and, most significantly, the ability to run custom software. Another essential feature of a smartphone is its user interface. A smartphone includes a touch screen and the ability to zoom in and out using basic interface buttons, menus, and forms, as well as the support of a 'qwerty' keyboard, making it straightforward to use for those who are unfamiliar with ICT technology and even not having enough educated. The program must be simple to use and the farmer must give only the information needed to complete an operation or procedure. The price of a smartphone ranges between low and expensive. As a result, farmers may easily purchase any type of smartphone that fits within their budget.

Smartphones and IoT are interrelated to each other. As a result, it is important in smart agriculture. Farmers can easily afford smartphones now that they are more affordable on the market. Furthermore, its computational capability allows users to develop a diverse range of useful apps. The android mobile application, also known as the android app, allows the user to monitor and operate the field from any location. An internet connection is required to monitor and operate the field. Rural farmers who formerly had limited access to up-to-date agricultural information (e.g., market, weather, and crop disease news) and assistance from agricultural experts and government extension workers now have new options, thanks to low-cost smartphones with a variety of sensors on the market. Even though all information is available in the public domain, accessing it is a time-consuming job for farmers. Mobile or smartphone applications may provide all of this information with changing seasons and climate. One of the platforms where a farmer may obtain all solutions and information in one touch is the mobile app. Farmers can use their smartphones to monitor their equipment, crops, and livestock remotely, as well as to obtain information on livestock nutrition and productivity. They may even utilize this technology to provide statistical forecasts for their crops and cattle [18]. Smartphone apps revolutionized connectivity and are used for sending agri-information to farmers. Farmers will also be notified via smartphone if an emergency arises on their farms.

9 Android Architecture

Android is an open-source development platform that allows developers to create extremely sophisticated applications. The Android operating system is made up of several software components that are grouped into five categories: Applications, Application Framework, Libraries, Android Runtime, Linux Kernel, and four major layers. It allows developers to take full advantage of the device's hardware, access location data, perform background services, call divert, and send SMS messages, among other things. After the program has been published, it can be downloaded through third-party websites or from online shops such as Android Market and the application store of Google [19]. Some of the android-based mobile apps for agriculture are:

- Agri App: Smart Farming App for Indian Agriculture.
- Kisan Suvidha.
- Damini.
- KVK Sandesh.
- MKisan Application.
- Agrimarket.
- Pashu Poshan.

10 Challenges in Implementing IoT

Farmers residing in rural areas are unable to take the benefit from IoT technology due to a lack of network infrastructure. As a result, they have limited access to the internet. A farmer needs consistent access to agricultural data at all times and from any place, thus a faulty connection renders an advanced monitoring system worthless. In addition, the types of equipment required to implementing IoT systems in agriculture are costly. Further, most people in this region still feel agriculture belong to their forefather generation; therefore they are hesitant to enter this industry. However, IoT should be brought closer to the primary sector by combining with complementing instruments to develop more efficient products. In this context, commercials and on-air promotions about new technology might be beneficial. To achieve an aggregated output, information from one farm can be shared with other farms.

11 Conclusion

Although IoT in agriculture is still in its early stage in India, how we are adopting this technology gives us reason to be optimistic in this area. The use of IoT in this cold arid region of Ladakh which remains inaccessible for half of the year from the rest of the country and physically reaching farmers in inaccessible areas is a real problem that has very good prospects. The use of the different types of sensors and other artificial intelligence devices can help ineffective ways of controlling, monitoring, and managing the farms. This will make the farmers in the region improve quality, quantity, minimize risks and wastes thereby making food items more accessible to customers and saving farmers time and money. Smart farming technology will also reduce the environmental effect of farming in this fragile region, ensuring long-term sustainability. What is required is to impart training to the farmers about these technologies so that they can perform their agriculture tasks quite easily without even reaching their field.

References

1. Shrivastava, R., Sharma, V., Jaiswal, V., Raj, S.: A research paper on smart agriculture using IoT. Int. Res. J. Eng. Technol. (IRJET) **7**(7), 2708–2710 (2020)
2. Sen, S., Madhu, B.: Smart agriculture—a bliss to farmers. Int. J. Eng. Sci. Res. Technol. 197–202 (2017)
3. District Statistical Handbook. Ladakh Autonomous Hill Development Council, Leh (2019)
4. Annual Progress Report. Krishi Vigyan Kendra Leh, Sher-e-Kashmir University of Agricultural Sciences & Technology of Kashmir (2020)
5. Mathad, A.S., Margaur, A., Lamani, B.: Smart agriculture system. Int. J. Sci. Technol. Eng. **4**(10), 75–77 (2018)

6. Lytos, A., Lagkas, T., Sarigiannidis, P., Zervakise, M., Livanose, G.: Towards smart farming: systems, frameworks, and exploitation of multiple sources. COMputer NETworks (COMNET) **172**, 1–14 (2020)
7. Jatana, R., Soni, P.: A study of smart farming agro-techniques and their applications in organic agriculture. Int. J. Creat. Res. Thought (IJCRT) **8**(8), 2956–2968 (2018)
8. Praveen, N., Ashif, A., Aleem, A.: IOT based automatic vehicle accident alert system. In: IEEE 5th International Conference on Computing Communication and Automation, pp. 330–333 (2020)
9. Aleem, A.: A concise artificial neural network in data mining. Int. J. Res. Eng. Appl. Sci. **2**(2), 418–428 (2012)
10. Nisha, B.N., Nandini, C., Pavithra, D.T., Anjum, R.: Survey on smart greenhouse monitoring based on internet of things. Int. Res. J. Eng. Technol. (IRJET) **7**(7), 397–399 (2020)
11. Mudholka, M., Mudholkar, P.: Study of improvement in efficiency and effectiveness of performing farming using internet of things. Int. J. Sci. Res. Dev. **5**(10) (2017)
12. Kamienski, C., Soininen, J.-P., Taumberge, M., et al.: Smart Water Management Platform: IoT-Based Precision Irrigation for Agriculture, vol. 19, no. 02. Licensee MDPI, Basel
13. Channe, H., Kothari, S., Kadam, D.: Multidisciplinary model for smart agriculture using internet-of-things (IoT), sensors, cloud-computing, mobile-computing & big-data analysis. Int. J. Comput. Technol. Appl. **6**(3), 374–382 (2018)
14. Mittal, S., Gandhi, S., Tripathi, G.: Socio-economic Impact of Mobile Phones on Indian Agriculture, p. 53. Indian Council for Research on International Economic Relations, New Delhi
15. Mishra, G.P., Singh, N., Kumar, H., Singh, S.B.: Protected cultivation for food and nutritional security at Ladakh. Def. Sci. J. **61**(02), 219–225 (2010)
16. Balakrishna, G., Rao Moparthi, N.: Study report on Indian agriculture with IoT. Int. J. Electr. Comput. Eng. (IJECE) **10**(03), 2322–2328 (2020)
17. Akshaya, M., Kavipriya, P.R., Yogapriya, M., Karthikamani, R.: IoT Based Smart Fertilizer Injector for Agriculture Plants, vol. 07, no. 03, pp. 2950–2954 (2020)
18. Barah, A., Malakrishnan, M.: Smart Phone Application: Role in Agri-information Dissemination, vol. 39, no. 01, pp. 82–85. Agriculture Research Communication Centre (2017)
19. Patel, H., Patel, D.: Survey of android apps for agriculture sector. Int. J. Inf. Sci. Tech. 61–67 (2016)

Integration of Artificial Intelligence and IoT on Agricultural Applications

Mantripragada Yaswanth Bhanu Murthy, Sk. Enaul Haq, Koteswararao Anne, and M. Sunil Babu

1 Introduction

The global food production is extended to rise by 70% by 2050 (according to "Food and Agriculture Organization of the United Nations n.d."). Indian farmers suffer from a lot of difficulties like water management, different crop diseases, natural disasters, climatic changes and lack of knowledge regarding their soil that degrades crop productivity rate. As the population of world increases gradually, the demand for the food production also increases in the agricultural field. Therefore, the artificial intelligence, machine learning, IoT and deep learning are used in the agricultural area for maximizing the food production with less manpower by detecting the crop diseases at an early stage and also by managing the soil and water for crop yield. Now-a-days, IoT and deep learning [1] are majorly used in the agriculture due to the automatic prediction of climatic changes and sensor-based applications. The IoT sensors facilitate several features to enhance the agriculture, which is depicted in Fig. 1.

IoT performs well in several other applications like security, traffic control system, health, transportation, supply chain, smart city and so on [2]. IoT is also utilized in agricultural areas for enhancing the quality of crop production outcomes. IoT sensor

M. Yaswanth Bhanu Murthy (✉) · Sk. Enaul Haq
Vasireddy Venkatadri Institute of Technology, Namburu, Guntur, Andhra Pradesh 22508, India
e-mail: mybhanu@vvit.net

Sk. Enaul Haq
e-mail: haq.enaul11@vvit.net

K. Anne
Kalasalingam Academy of Research and Education, Srivilliputhur, Virudhunagar, Tamil Nadu 626128, India

M. Sunil Babu
K.L.E.F, Vaddeswaram, Guntur, Andhra Pradesh 522502, India

S. Mukherjee et al. (eds.), *Intelligent Systems for Social Good*,
Advanced Technologies and Societal Change,
https://doi.org/10.1007/978-981-19-0770-8_3

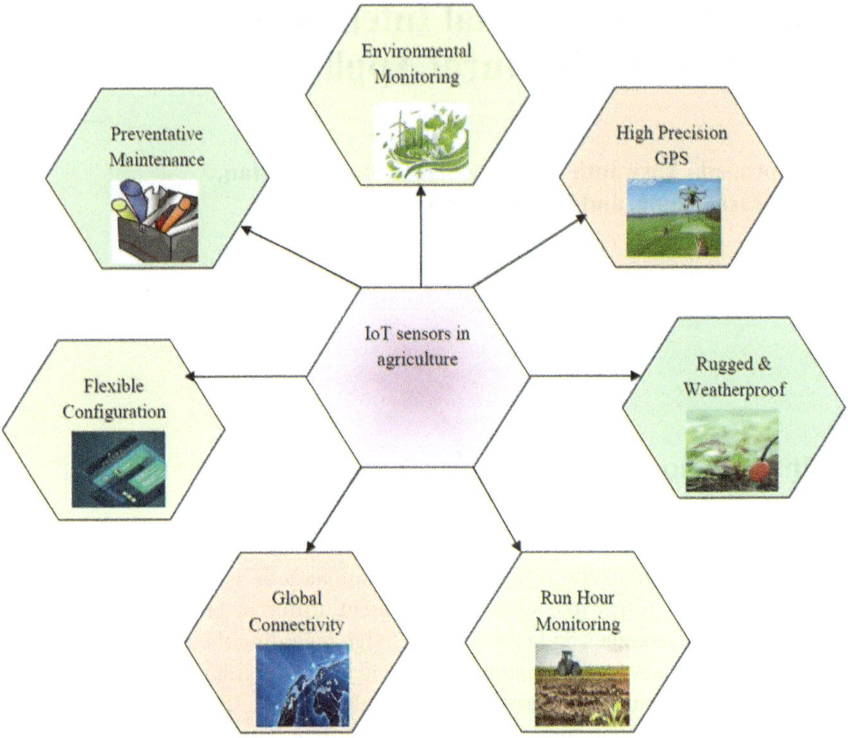

Fig. 1 IoT sensor features in agriculture

devices provide numerous amounts of data that may be unstructured, semi-structured and structured. Further, these data will undergo analytics for predicting the future and determining the present information. Analyzing these data is complex because of its huge quantity. Anyway, these data are much essential when it produces an action and also it can be possible to process with a smart learning mechanism. The main attention is to identify the way for analyzing huge amount of information from the complex data of IoT [3]. Here, the deep learning technology is utilized and also employed in different IoT applications due to its capacity on achieving the efficient results in shorter time. It is observed that the deep learning approaches are much efficient in resolving the problems related to real-time applications.

The huge amount of data are collected mostly in the form of images through remote sensing, in which the collected images represent numerous challenges in the agricultural environments. The agricultural domain is mainly dependent on the imaging analysis for identifying and detecting the anomaly through these images, and also the intelligent data analytics is also applied for different applications of agricultural areas [4]. The major objective of this chapter is to determine the applicability of the deep learning techniques in the agricultural field along with the IoT for enhancing the capacity of the sensor-based applications. Moreover, machine learning and artificial intelligence are also enriched the various agricultural domains such as analytics of environmental prediction, automated agricultural growth, precision farming, etc. [5].

2 IoT and Artificial Intelligence in Agriculture

IoT plays a major role in different real-world applications. Similarly, a new framework is introduced in the smart agriculture by employing different sensors such as "temperature sensor, water level sensor and soil moisture sensor" for the purpose of observing the agricultural activities like plant, crop and irrigation monitoring. So, it is easy for the farmers to monitor the agricultural activities from any place. While comparing the ordinary farming like reaping, manual tilling, and planting with the IoT-based smart farming, it provides better performance to increase the agricultural productivity rate by utilizing several sensors in field. Smart agriculture is one among streams in the IoT [6]. It produces numerous data through the sensors that bring out the notion of big data, which is considered to be huge amount of data from different sources such as "sensor, social networking and business". The challenges are occurred during the data recording, storing, analyzing and searching the data from the IoT devices. The analytics process of agricultural data [7] is given in Fig. 2.

Additionally, these data were obtained from different varieties of sensor systems, and so, it has shown the heterogeneity nature in the attained information. This creates the primary challenges while analyzing the real-world IoT data that affect the retrieval of useful information. To overcome these issues related to IoT big data, enormous analytical solutions are introduced for determining the valuable information of big data through IoT devices. Smart farming based on IoT is essential in providing several benefits like offering appropriate fertilizer and controlling of water optimum usage [8]. The IoT applications in the agricultural field are described in Fig. 3.

3 Overview of Deep Learning

Deep learning works on the basis of computational frameworks or Artificial Neural Networks (ANNs), in which the functions of deep learning are similar with the working of the human brain. Some of the deep learning applications are driverless cars, voice control in phones, hands-free speakers and tablets, automatic game playing and also in natural language processing, where these models are trained with the images, text or speech for the classification. It includes a huge quantity of training data that are obtained from the neural networks with numerous hidden layers. Most of the deep learning methods include the neural networks for training the data, and so, it is termed as deep neural networks. The neural networks get the support of several learning algorithms for enhancing the information quantities, and thus, it provides the effective training process [9]. While utilizing the huge dataset, the processing of the model will be more efficient.

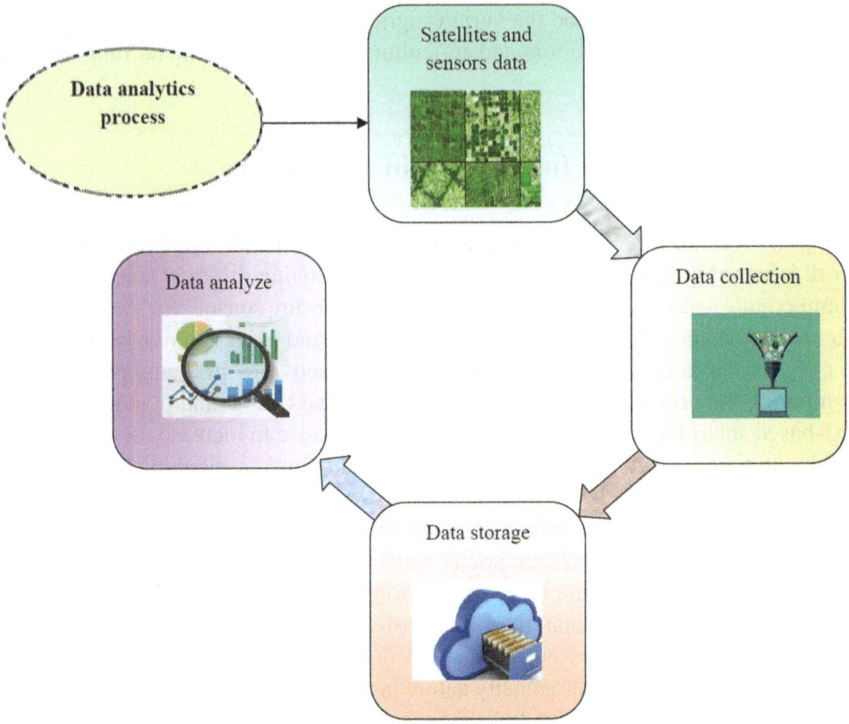

Fig. 2 Data analytics process in agriculture

Deep learning employs various algorithms for information processing and enhances the decision-making ability. This type of invention is utilized in certain sectors like recognition of pictures and voice and self-driving cars. It is more difficult to manage the complex and high-dimensional data in the nature language processing, image translation and image processing techniques. This also leads to complexity in processing these high-dimensional data. These problems related to processing the complex data can be overcome by the deep learning models by utilizing its conceal layer and also produce the subsequent information for predicting and detecting the farming tasks. The main advantage of deep learning takes place in the feature extraction, where the features can be extracted from the raw data and generate the accurate features. The feature extraction is the process of converting the input data into a set of features, which is further utilized for the classification or recognition purposes. Deep learning does not contain any requirements for physically extracting the image features [7]. At the time of training process, the network itself obtains the extracted features and put forwarded towards the network for processing without any manual support, and finally verifies the dataset. The process of deep learning in agriculture consists of gathering of data from sensors and other devices, analysis and extraction of features, make predictions, train, and test model, and makes predictions. These consecutive processes in deep learning approaches help in learning and boost the

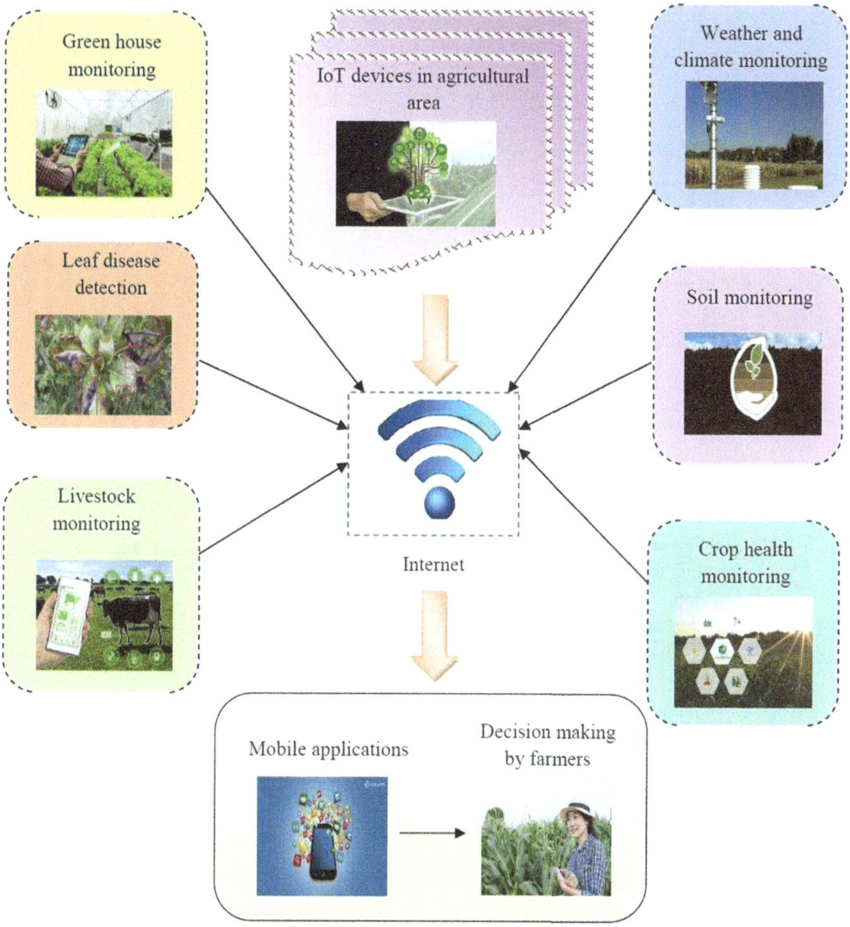

Fig. 3 IoT applications in agriculture

analytics in IoT applications. The deep learning process with IoT sensors for making prediction about the agricultural areas is diagrammatically represented in Fig. 4.

4 Deep Learning for Smart Agriculture

Deep learning is much essential in the "automation of predictive analytics". It is useful in certain agricultural areas such as hydroponic agriculture, detection of fruit type, crop or plant classification and fruit counting. In agriculture, it is necessary to have healthy crops for enhancing the quality of the agricultural production. Deep learning [10] has ensured its efficiency in solving the problems like image segmentation, image

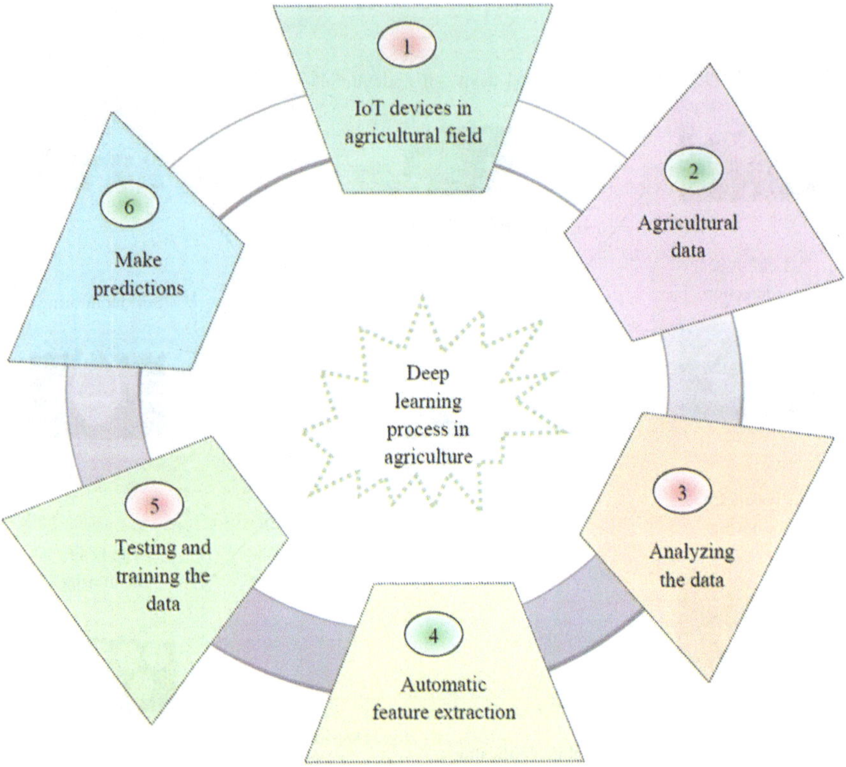

Fig. 4 Deep learning process using IoT sensors in agriculture

classification, object detection, picture recognition and natural language processing. It provides the enhanced accuracy for the classification even in processing the huge quantity of dataset. It is also useful in various functions such as automatic farming detection and forecast.

5 Several Deep Learning Techniques in Agriculture

Different agricultural applications in IoT are detection of plant diseases, forecast of soil moisture content, fruit counting, detection of leaf diseases, crop type classification and plant recognition [11]. To support agriculture, healthy crops should be produced, and so, artificial intelligence plays a major role in solving the challenges like image segmentation and classification, object detection. The existing studies show that the superior classification accuracy is observed by deep learning approaches. Some of the common deep learning approaches are "Convolutional

Neural-Network (CNN), Recurrent Neural Networks (RNN), Generative Adversarial Networks (GAN), Long- Short Term Memory (LSTM)", etc.

Convolutional Neural-Network: It is a set of feed-forward and deep ANN, which is utilized for analyzing the visual images. This network contains numerous numbers of neurons with learnable weights and biases. The input layer is inserted with the input data, which is further given to the several hidden layers, and results are obtained from the output layer. While comparing with other ANN models, CNN has convolutional and pooling layers to extract the features. Here, the convolutional layers perform the feature extraction using the input images, in which the dimensionality of the features will be minimized by the pooling layer. It has the capacity to learn the complex problems at a faster rate because of its complex and weight-sharing model that enables the parallelization. When the large datasets are provided to the CNN, it achieves more accuracy in the classification. So, data augmentation is used to increase the size of the dataset and enhance the CNN accuracy [12]. CNN is independent of illumination, size or translations, and so, continuous random transformations like translation and rotation can be applied to input images for enhancing the dataset. This is more useful in the different areas of farming fields such as "crop and plant leaf disease detection, plant recognition, land cover classification, weed identification, and fruit counting".

Recurrent Neural Networks: It is said to be a multi-temporal network that contains network connections among the direct graphs and its nodes with the temporal sequence. RNN has an ability to allow the temporal dynamic behavior. It has the capacity to utilize the information present in its own internal memory. This indicates that RNN employs previous observation for obtaining the classification output of the current observation.

Generative Adversarial Networks (GAN): GANs are designed by considering two neural network systems like discriminative networks and generative networks, which help in inspecting, interpreting and mimicking from the training dataset. It can be enhanced by using GAN. The combined GAN architecture gives high-quality data.

Long-Short Term Memory (LSTM): It is one of the efficient algorithms, among other deep learning techniques, which process complete data and also process the single information points. Therefore, LSTM is useful for efficient forecasting and classification of time-series data. The LSTM is more eminent in processing agricultural applications.

6 Frameworks in Deep Learning

Certain frameworks of deep learning are enclosed in the agricultural field, such as "Convolution Architecture for Feature Extraction (CAFFE), PyTorch, Theano and TensorFlow".

CAFFE (Convolution Architecture for Feature Extraction)

It is an "open source framework" that ensures different types of deep learning architectures. This framework function utilizes the C++, which elevates the "Compute Unified Device Architecture (CUDA)" for computing the "Graphics Processing Units (GPU)". This ensures the interface for python and MATLAB. This framework has supported numerous deep learning methods for processing image segmentation and classification. This also enforces the fully connected neural network designs such as LSTM and CNN.

PyTorch

It is an "open-source code framework" that helps the machine learning algorithms. It is developed by Torch, which is a scientific computing framework that is utilized for developing and training the deep neural network. It is simple to make the DNN models when using the PyTorch framework. This framework has numerous deep learning algorithms that are constructed based on PyTorch. Due to its clear architectural style, it is easy to process the training and implementing the deep learning for the purpose of learning and execution.

Theano

Theano is "an open source Python language" that works based on the deep and machine learning framework. Theano ensures fast computation, and it can be run on both GPU and CPU. But, it works slower on CPU while comparing the processing time of GPU. It is used for training the deep neural network algorithms.

Tensor Flow

An "open source software library" is Tensor Flow that is used for machine learning and deep learning and also for different deep neural networks. It is very essential for generating the neural networks through the graph representations. Here, the nodes indicate the mathematical operations performed based on the graph, and edges denote the "multi-dimensional data arrays (tensors)" that is located in between the nodes. A set of imaging tools named TensorBoard is implemented by TensorFlow for visualizing the TensorFlow outcomes.

7 Deep Learning Applications in Agriculture

In recent times, IoT and robotics technologies, computer vision, deep learning and image processing are very useful for farmers. These methods are used for visualizing the progress of the crops, which can support the farmers to determine the status of crops whether it can be moved to harvest or not. The deep learning applications in agriculture are unlimited, where some of them are described as follows.

Crop Management: It is the most essential part of agriculture for enhancing the crop quality. The deep learning techniques are utilized in the smart agriculture for observing the water level and temperature of the crop. Moreover, the farmers can monitor their agricultural land from other areas. Smart agriculture can be effective while performing it with the IoT and artificial intelligence [13].

Water Management: The crop growth depends on the water level, and so, analyzing the water level regularly is much significant for improving the growth of the plant. By considering the soil and crop category, it is very effective in machine learning algorithms for monitoring the level of the water [14]. Further, the artificial neural network model is developed for "smart irrigation system" that ensures the enhanced water management in the agricultural areas.

Soil Management: It is easy to identify the appropriate crops and its suitable fertilizer for the soil by utilizing the machine learning methods. The soil quality is based on its macro, micro and primary nutrients and pH level, which are examined for estimating the soil health condition [15]. The soil management is enhanced by introducing the "artificial neural networks, Bayesian network and Gaussian kernel-based support vector machines".

8 Conclusion

As the population of the world increases every day, the demand for food production also increases. Hence, the food production can be improved by assisting the farmers. The new research deploys IoT devices along with artificial intelligence for improving the investments. The smart farming is improved by adopting the artificial intelligence with IoT for efficient processing and generation of information. The sensors or IoT devices can be installed in IoT applications for making the agriculture efficient one. The gathered information is processed by deep learning approaches, which assists the farmers in the future for smart farming. This chapter has explained the contributions of IoT and artificial intelligence, deep learning in the smart agriculture. Also, some deep learning approaches and frameworks for improving the field of agriculture were explained and also have discussed about the applications of deep learning in agriculture. Artificial intelligence or IoT in agriculture causes the cost inefficiency for buying the robots and requires regular maintenance for its effective working. It also affects the poor farmers due to its high energy cost and maintenance. So, in future, it is necessary to develop the robots at an affordable rate for assisting the farmers.

References

1. Hu, W., Fan, J., Du, Y., Li, B., Xiong, N., Bekkering, E.: MDFC–ResNet: an agricultural IoT system to accurately recognize crop diseases. IEEE Access **8**, 115287–115298 (2020)
2. Horng, G., Liu, M., Chen, C.: The smart image recognition mechanism for crop harvesting system in intelligent agriculture. IEEE Sens. J. **20**(5), 2766–2781 (2020)

3. Lammie, C., Olsen, A., Carrick, T., Azghadi, M.R.: Low-power and high-speed deep FPGA inference engines for weed classification at the edge. IEEE Access **7**, 51171–51184 (2019)
4. Chen, C.–J., Huang, Y.–Y., Li, Y.–S., Chang, C.-Y., Huang, Y.-M.: An AIoT based smart agricultural system for pests detection. IEEE Access **8**, 180750–180761 (2020)
5. Shafi, U.U., Mumtaz, R., Iqbal, N., Zaidi, S.M.H., Zaidi, S.A.R., Hussain, I., Zahid, M.: A multi-modal approach for crop health mapping using low altitude remote sensing, internet of things (IoT) and machine learning. IEEE Access **8**, 112708–112724 (2020)
6. Mishra, M., Choudhury, P., Pati, B.: Modified ride-NN optimizer for the IoT based plant disease detection. J. Ambient Intell. Hum. Comput **12**, 691–703 (2021)
7. Wani, J.A., Sharma, S., Muzamil, M., Ahmed, S., Sharma, S., Singh, S.: Machine learning and deep learning based computational techniques in automatic agricultural diseases detection: methodologies, applications, and challenges. Arch. Comput. Methods Eng. (2021)
8. Tan, W., Zhao, C., Wu, H.: Intelligent alerting for fruit-melon lesion image based on momentum deep learning. Multimed. Tools Appl. **75**, 16741–16761 (2016)
9. Song, X., Zhang, G., Liu, F., Li, D., Zhao, Y., Yang, J.: Modeling spatio-temporal distribution of soil moisture by deep learning-based cellular automata model. J. Arid Land **8**, 734–748 (2016)
10. Khan, S., Tufail, M., Khan, M.T., Khan, Z.A., Anwar, S.: Deep learning-based identification system of weeds and crops in strawberry and pea fields for a precision agriculture sprayer. Precis. Agric. (2021)
11. Kong, J., Wang, H., Wang, X., Jin, X., Fang, X., Lin, S.: Multi-stream hybrid architecture based on cross-level fusion strategy for fine-grained crop species recognition in precision agriculture. Comput. Electron. Agric. **185**(106134) (2021)
12. Lu, H., Fu, X., Liu, C., Li, L.-G., He, Y.-X., Li, N.-W.: Cultivated land information extraction in UAV imagery based on deep convolutional neural network and transfer learning. J. Mountain Sci. **14**, 731–741 (2017)
13. Chehri, A., Chaibi, H., Saadane, R., Hakem, N., Wahbi, M.: A framework of optimizing the deployment of IoT for precision agriculture industry. Procedia Comput. Sci. **176**, 2414–2422 (2020)
14. Vij, A., Vijendra, S., Jain, A., Bajaj, S., Bassi, A., Sharma, A.: IoT and machine learning approaches for automation of farm irrigation system. Procedia Comput. Sci. **167**, 1250–1257 (2020)
15. Bu, F., Wang, X.: A smart agriculture IoT system based on deep reinforcement learning. Future Gener. Comput. Syst. **99**, 500–507 (2019)

Internet of Things-Based Smart Transportation System for Smart Cities

E. Fantin Irudaya Raj and M. Appadurai

1 Introduction

Many developing countries like India invest a huge amount of money to transform their old existing cities into modern smart cities. A smart transportation system is an integral part of this planning and commissioning. Advancement in the Internet of Things (IoT) and Information and Communication Technology (ICT) plays a crucial role in smart city development. The smart city concept combines ICT and IoT to improve the efficiency of city services while also connecting them to inhabitants. It facilitates communication between officials and the public, as well as civic infrastructure, in order to monitor cities for various objectives. The quality and interactivity of city services are expected to augment with the help of IoT.

These technologies also play a vital role in a modern transportation system. Vehicles with sophisticated accident-prevention systems, such as collision warning systems (CWSs) or lane-keeping assistance (LKA), are now available in the market. The next stage in decreasing road accidents is to plan ahead of time for such cars to minimize collisions while also improving traffic flow. In order to effectively manage traffic problems, vehicle-to-infrastructure (V2I) communications are required.

Self-driving cars or autonomous vehicle is also a part of this modern transportation system. Many researchers are working towards this area in recent times. The Internet of Things (IoT), Information and Communication Technology (ICT), remote sensing, Global Positioning Systems (GPS), and embedded systems all contribute

E. Fantin Irudaya Raj (✉)
Department of Electrical and Electronics Engineering, Dr. Sivanthi Aditanar College of Engineering, Tiruchendur, Tamil Nadu, India
e-mail: fantinraj@gmail.com

M. Appadurai
Department of Mechanical Engineering, Dr. Sivanthi Aditanar College of Engineering, Tiruchendur, Tamil Nadu, India

S. Mukherjee et al. (eds.), *Intelligent Systems for Social Good*,
Advanced Technologies and Societal Change,
https://doi.org/10.1007/978-981-19-0770-8_4

to the development of this autonomous vehicle technology. Advancements in these areas make the autonomous vehicle's maneuvering easier and more precise. Electric vehicle adoption and converting the existing gasoline vehicles into electric vehicles also play a crucial role in the future transportation system. The smart cities also needed to plan accordingly and make proper arrangements for vehicle charging and all.

2 Literature Review

The last century saw substantial improvements in the quality of life, particularly in terms of access to services. However, administrators, architects, and urban planners have faced a significant problem due to intensive industry and the rising population in metropolitan regions [1]. Over the last decade, the Internet of Things (IoT) and information and communication technologies (ICT) have had a tremendous impact on how businesses approach innovation and how they originate and exploit opportunities in their daily operations. These were compounded in Smart Cities, where the purpose of the IoT is to leverage ICT to allow value-added services for citizens while also giving enterprises more opportunity to innovate by integrating cutting-edge technology [2]. One facet of the smart city is smart transportation. Transportation has become the second-largest source of carbon emissions due to its low efficiency. It has an impact not only on smart transportation but also on the smart environment. As a result, improving transportation efficiency is critical to smart transportation and smart cities [3]. Although transportation has substantially improved our lives, numerous major challenges remain unsolved, including auto crashes, traffic jams, and vehicle discharge. Intelligent transportation has recently become a hot topic in the Internet of Things (IoT) space, and it is seen as a solution to the challenges mentioned above [4].

Applications have become smarter as the Internet of Things has grown in popularity, and connected gadgets are being used in all areas of modern cities. Machine Learning (ML) approaches are used to improve the intelligence and capabilities of an application as the volume of acquired data is expanding [5]. As the world's population grows, so does the number of vehicles on the road, increasing traffic management issues, particularly for public transportation. Furthermore, the frequency of accidents and other traffic-related problems is increasing. The Intelligent Transportation System (ITS) overcomes most of these problems by merging existing technologies with the fundamental infrastructure [6]. Because of mobile technology and the ubiquity of the cellular network, real-time vehicle tracking for effective transportation management is now conceivable. The Intelligent Public Transportation System eliminates the need to wait for hours for a bus. Smartphones are one of the most enticing alternatives for developing IoT apps due to their omnipresent nature and ever-increasing power at a low cost. A system based on a blend of technologies such as GPS and Android is being investigated here to reassure consumers who utilize public transit [7].

A decentralized data management system for smart and safe transportation is built to tackle the data vulnerability problem, leveraging blockchain and the Internet of Things in a sustainable smart city framework [8]. Electric Vehicles (EVs) are predicted to become widely used in personal, business, and public fleets in urban cities in the future. The popularity of electric vehicles will greatly impact the city's long-term sustainability and economic prosperity [9]. The effective management of hybrid electric vehicles with modern machinery and its effective way to control the machine is also discussed [10–12]. The various electric vehicles can be adopted for smart city environments, and their charging methodologies are also explained [13–15].

Vehicle parking is also another major concern in many smart cities. Because most of today's smart cities were converted from older traditional cities, they may have limited roads that can't handle a huge number of vehicles. As a result, finding a parking spot at the right time and in the right location has become an inevitable requirement for people worldwide. However, finding the right place to park the vehicles may also lead to time wastage and more gasoline consumption. The availability of parking occupancy information in advance plays a critical role in ensuring that motorists have a stress-free trip. It also makes service-profit management easier for parking lot owners. It also aids in the control of pollution in smart cities by lowering traffic congestion by reducing cruising time [16, 17].

Smart transportation design and planning, smart communities, smart cities, smart control systems, and other sectors all use big data analytics. A model for evaluating transportation data utilizing Hadoop and Spark is proposed to handle real-time transportation data. The suggested system is put to the test using transportation data from a variety of reliable sources. The results reveal data processing and real-time distribution to citizens in the shortest time possible [18–20]. Vehicle-to-vehicle communication and Vehicle-to-Infrastructure communication are playing an important role in smart transportation [21, 22]. These techniques are creating the basic foundation for autonomous vehicles for future smart cities.

From the review of the literature, we can understand the importance of IoT-based smart transportation systems. Further, the following sections explain the IoT-based Traffic Management system, vehicle parking system in detail. In addition, the autonomous vehicles for future smart cities in the modern transportation system are also explained in detail.

3 Autonomous Vehicles for Smart Cities

In recent years, the smart city project has been implemented in urban areas to enhance the sophisticated life of the people by adopting advanced technology. The people are interested in utilizing the modern innovations in network connectivity, garbage handling, transportation, traffic management, city monitoring, security, water management, and autonomous vehicles. Especially in the automotive field, the three modern technologies such as electric vehicles, connected vehicles, and autonomous vehicles are revolutionized. The electric vehicle gives the option for a pollution-free environment inside the urban region. The autonomous vehicle provides crash less driving and controls the rash driving of youngsters. The connected vehicle has an internet connection that gets onsite information about the city traffic and navigation details from the cloud storage. These benefits are employed by establishing smart devices throughout the city. However, without these sensors, transducers, and high-speed wireless internet connectivity, these modern innovations can't be utilized by people.

The city traffic can be efficiently controlled with the help of autonomous vehicles by taking dynamic decisions in traffic management from real-time information. For an efficient traffic management system, the smart city project needs base technology such as self-driving technology, vehicle-to-vehicle communication, vehicle-to-infra communication, and smart parking system. These concepts and their needs are briefly discussed in the below sections.

With the implementation of driverless vehicles in smart cities, people get several benefits. Road accidents and transportation costs are entirely reduced. The traffic congestion is efficiently managed. The optimum driving speed inside the city can get implemented. The vacancy in parking spaces is given as on-board information, so the time to search the parking area is reduced. Thus, the fuel economy is improved by reducing unnecessary parking area searches and city traffic, and thus, the greenhouse gas emissions can significantly decrease. On the road, there is a provision that gives priority to emergency vehicles and medical waggons. The autonomous vehicle provides overall public safety in roadways.

3.1 Vehicle-To-Vehicle Communication

Vehicle-to-vehicle communication empowers the automotive to exchange their information among the vehicles. The speed, destination, and navigation are wirelessly broadcasted to the nearby vehicles. The proper software tools are installed in each vehicle to exchange the data without potential threats. The driver is alerted by messages and warning alerts for necessary control over the vehicle movement to avoid accidents. The message transfer is done more than 300 m before to alert drivers regarding the traffic, weather conditions, threats, and general information. With the assistance of vehicle-to-vehicle communication and roadside sensors, road accidents are minimized to enhance road safety. The best sample for this safety method is adaptive cruise control technology [23], in that sensors are utilized to avoid vehicle collisions. An automotive vehicle integrates with an adaptive control device. Therefore, the speed is spontaneously regulated to move at a particular speed. It is implemented with the assistance of vehicle-to-vehicle communication only. As a result, the automotive is efficiently coordinated in a specific region by controlling the speed and applying brakes automatically on demand. The dynamic clustering of autonomous vehicles improves road lane utilization and reduces inner-city traffic

[24]. For this, efficient data analysis techniques are essential. Since the sensor information is not always examined. The required onsite information is only monitored for dynamic analysis.

IoT devices assist in coordinating the several devices to connect with embedded computing tools. The IoT connects with the sensors by wireless network connection. The smart city project utilizing IoT tools has various benefits, such as being user-oriented and better performance. The several sensing elements like alcohol sensor, vibration measuring transducer, accelerometer, and pressure transducer are linked with IoT devices to avoid highway accidents. The device enhances driver safety by getting simultaneous data from the sensor nodes to avoid crashes among vehicles. The standard procedure is followed during the lane change [25] in an autonomous vehicle to avoid a collision. The vehicle is moved to another lane when the required spacing is available. The driverless vehicle must maintain adequate space among the host and surrounding vehicle to avoid a crash.

The heavy freight vehicle consumes a large amount of fuel for its displacement. Fuel saving is a universal requirement due to the increase in fossil fuel prices and Greenhouse gas emissions. The sensors in the vehicle-to-vehicle communication concept give better solutions for fuel-saving [26]. The sensors in the automotive analyze geological data such as wind velocity, road elevation for dynamic decision making. The sensor communicates with the infrastructure to know about city traffic and possible clear routes. The vehicle-to-vehicle communication method gathered the data beyond the eyesight, and acquired information is used for efficient traffic management. The feasible route is identified and communicated with the vehicle handler. Thus, vehicle fuel saving is improved by this concept. Figure 1 illustrates

Fig. 1 Vehicle-to-vehicle communication

the detailing of vehicle-to-vehicle communication.

Sustainable development is the main objective of modern technologies, and the smart city project creates a way for people to live in the available environment in comfort. The balancing of the resources in digital cities is enhanced by the proper handling of all available resources eventually. For example, public transport is viable to all types of people by providing separate expressways. The time delay of the public transport due to city traffic in the highway is eradicated by this special expressway. The daily workers use the public vehicles for their routine trips, and this reduces separate vehicle usage. The smart sensors interact with the vehicles and infrastructure to monitor by analyzing gathered information. The smart city design is capable of handling the more challenging sustainable projects inside congested cities. The supply chain management related to electricity supply, garbage management, material consumption, and wastewater recycling is done with the assistance of smart sensors [27].

3.2 Vehicle to Infra Communication

A smart city utilizes several electronic devices and sensors to gather information from several nodes. The transportation system can communicate with different infrastructures [28] available in the digital city project. The social transformation is happening by the smart city project and fifth-generation network connection. The 5G network implementation [29] depends on the population and economic wealth of the nation. The automobile sector is also moving towards the utilization of higher-end technology. The modern connected vehicles receive and transmit information from their environment. The rescue vehicle or hospital wagon accesses the roadways easier by smart technology from such information. The road traffic is efficiently managed. Traffic congestion, pollution, land utility, energy demand are reduced, and mobility and safety are increased from modern innovations in connected vehicles. The data exchange between vehicles or between vehicles and infrastructure are essential in smart traffic management projects. The location, velocity, and navigation of the vehicle are continuously monitored to prevent accidents. The dynamic routing is done on heavy traffic by the gathered data from the vehicle to infra communication. Figure 2 shows the vehicle to infrastructure communication intangible diagram.

The connected vehicles have in-built sensors capable of analyzing their surroundings without human interference while navigating. Automated vehicles run on the road due to the advancement in network connections and computing technology. Modern communication technologies like 4G and 5G allow driverless vehicles to connect with thousands of network nodes, neighboring cars, and roadside electronic devices.

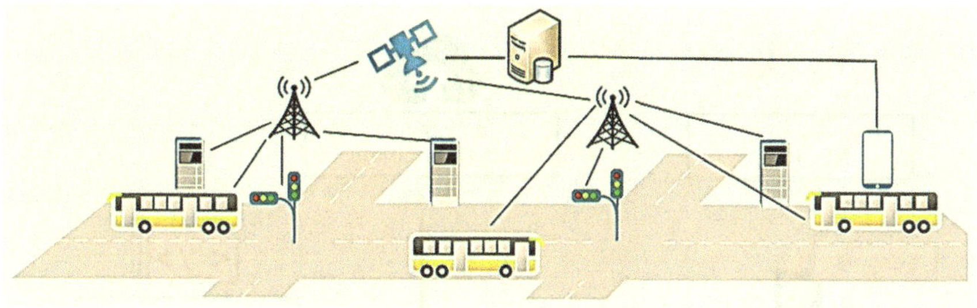

Fig. 2 Vehicle to infra communication

3.3 Vehicles Platooning

The navigation of vehicles in tightly aligned units on the highway roads is called vehicle platooning. This concept gives several advantages: enhanced traffic management, higher road capacity, crashless driving, better fuel economy, and idle movement. The vehicle handlers get relaxed without driving and do other work. It increased the productivity of the human workforce by doing other tasks on traveling. For the platoon vehicles, continuous information transfer with nearby automotive is supported. Thus, vehicle-to-vehicle communication is very important in vehicle platooning. The joining or relieving of vehicles is done at any time on active platoon movement [30]. The alerting of nearby automotive is made on platoon forming and leaving. This alert assists the vehicle to join or smoothly leave the platoon without disturbing other vehicles. Messages indicating acceleration, navigation, routing, platoon head change, and braking are transmitted to the vehicle to ensure safety of the vehicle.

3.4 Intelligent Transportation System

The sensors, IoT devices, and wireless networks are efficiently used in data exchange on intelligent transportation systems in smart cities. The vehicles have sensors for gathering onsite data. The dynamic information about vehicle location, navigation, and speed is provided by short message type services to the drivers. The on-road movement of automotive is made to the required destinations without human interference. Smartphones connected with vehicle networks give real-time information. The roadside sensor is erected at a uniform distance to gather data about the vehicle movement in that region [31]. Figure 3 shows the conceptual diagram of roadside sensor installation. The IoT devices can develop interconnectivity with the surrounding devices automatically.

The vehicle needs to connect with another vehicle, the roadside acts as the intermediary device among them. The roadside sensor provides an intermediate connection to

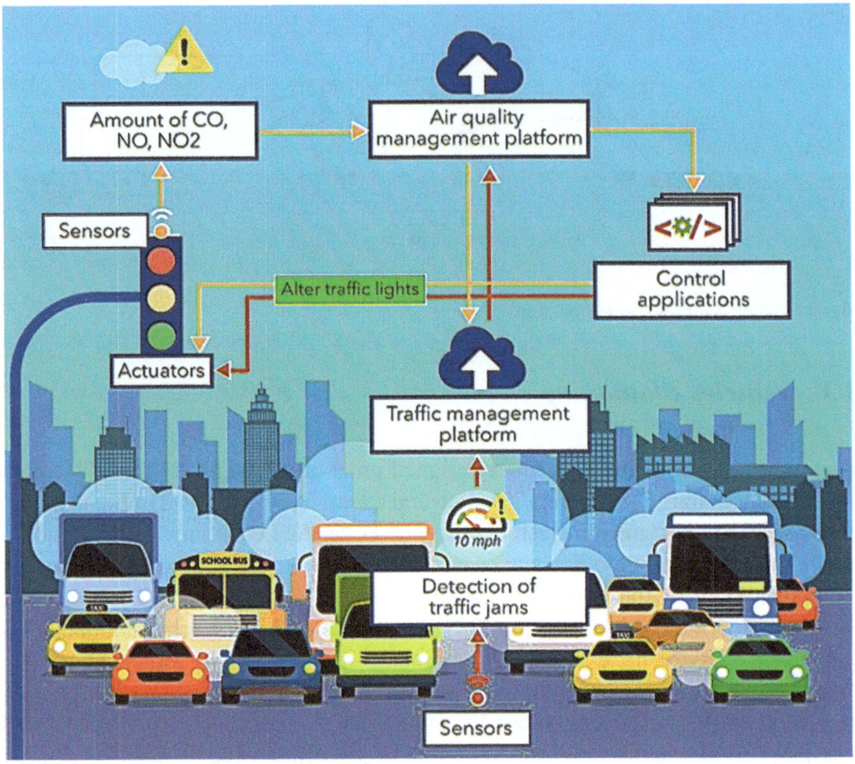

Fig. 3 Roadside sensors

a particular range. The alert having vehicle information is given to the owner on unauthorized access of the automotive. An intelligent transportation system effectively handles city traffic. The first step in the traffic management concept is registering the vehicle data to the roadside sensor. The destination of the vehicle to reach is also given. The roadside sensor in one region is interconnected to the other sensor to gather the traffic details. A first vehicle route is analyzed about the traffic details. If it is feasible, the first route is provided; otherwise, the next route is considered and so on. Thus, the optimum route is given to the vehicle for reducing traffic congestion in urban areas. IoT devices deliver the information with the optimum route via message. Implementing this technology makes it possible to create a traffic-free environment in modern cities. This technique provides the suitable and fastest route for forwarding hospital wagons. The safety, parking space handling is enhanced with the available traffic information [32]. The unnecessary time delay for travel is efficiently controlled, and the traffic problem is reduced.

4 IoT-Based Vehicle Parking System for Smart Cities

Normally, the cities are furnished with tall residential buildings, giant commercial malls, and densely populated vehicles. Each family has at least one vehicle for their transport. The special or collective parking area is developed in the apartments and commercial malls to facilitate their customers. The lift-type shaft type parking slot is also built-in densely populated cities. This considerable investment is opened correctly for the people's benefit. The investment amount gets payback only through proper utilization of resources. Electronic devices such as IoT are installed in each region to gather real-time information for traffic management.

The free parking slop searching is a huge task for the drivers in the inner city traffic. The time and fuel are wasted in searching the parking space. The connected vehicle has a sensor that provides information about the location and navigation details. The parking area is equipped with IoT to load onsite information to the central server. The central server gets data from several network nodes and delivers it as needed. People can use apps on their cell phones to obtain information from the cloud server. Before arriving at a specific location, the vehicle operator can look for available parking spaces. When the moving destination is input into the app on the smartphone, the central server sends numerous pieces of information. The essential details such as city traffic, weather condition, optimum route to move to avoid city traffic, parking slot free space, distance to travel, and the time to reach the destination are available in the smartphone app. The unnecessary searching of the parking area is avoided in this smart parking concept. The pollution and fuel wastage due to severe traffic and parking space examining is controlled. It offers cost-beneficial returns on smart project investments. The automatic movement of the car to the parking slot is depicted in Fig. 4.

Fig. 4 Smart parking

The autonomous vehicle further utilizes the driverless benefits. Only the smartphone app allows you to choose a parking spot. Self-driving technology moves the vehicle to and from the parking area without the need for human interaction. The smart vehicle concept is being developed further. When people travel to work in autonomous vehicles, there is no need for them to drive. They can do their work in the car at the time of traveling. After reaching the destination, the vehicle automatically moves to the parking space. The vehicle service and fuel refilling are also done by providing commands through a smartphone app.

5 Conclusion

The present work discussed IoT-based Smart and intelligent Transportation systems for smart cities. Autonomous vehicles for smart cities and their importance are also getting discussed. The underlying principles like vehicle-to-vehicle communication and vehicle-to-infrastructure communication of autonomous vehicle's technology are also explained. Another important issue in many smart cities is their vehicle parking system. The IoT-based modern system addresses this issue effectively and acts more precise and efficient than a conventional parking system. We conclude that implementing IoT-based smart transportation systems has overcome the daily issues faced by the inhabitants of smart cities because of transportation. As a result, people's lives in smart cities will improve, their productivity and effectiveness in the current environment will increase, and their entire facet will alter.

References

1. Eremia, M., Toma, L., Sanduleac, M.: The smart city concept in the 21st century. Proc. Eng. **181**, 12–19 (2017)
2. Bresciani, S., Ferraris, A., Del Giudice, M.: The management of organizational ambidexterity through alliances in a new context of analysis: Internet of Things (IoT) smart city projects. Technol. Forecast. Soc. Chang. **136**, 331–338 (2018)
3. Lingli, J.: Smart city, smart transportation: recommendations of the logistics platform construction. In: 2015 International Conference on Intelligent Transportation, Big Data and Smart City, pp. 729–732. IEEE (2015)
4. Satyakrishna, J., Sagar, R.K.: Analysis of smart city transportation using IoT. In: 2018 2nd International Conference on Inventive Systems and Control (ICISC), pp. 268–273. IEEE (2018)
5. Zantalis, F., Koulouras, G., Karabetsos, S., Kandris, D.: A review of machine learning and IoT in smart transportation. Future Internet **11**(4), 94 (2019)
6. Sutar, S.H., Koul, R., Suryavanshi, R.: Integration of smart phone and IOT for development of smart public transportation system. In: 2016 International Conference on Internet of Things and Applications (IOTA), pp. 73–78. IEEE (2016)
7. Raad, M.W., Deriche, M., Sheltami, T.: An IoT-based school bus and vehicle tracking system using RFID technology and mobile data networks. Arab. J. Sci. Eng. **46**(4), 3087–3097 (2021)

8. Abbas, K., Tawalbeh, L.A.A., Rafiq, A., Muthanna, A., Elgendy, I.A., El-Latif, A., Ahmed, A.: Convergence of blockchain and IoT for secure transportation systems in smart cities. Secur. Commun. Netw. (2021)
9. Cao, Y., Ahmad, N., Kaiwartya, O., Puturs, G., Khalid, M.: Intelligent transportation systems enabled ICT framework for electric vehicle charging in smart city. In: Handbook of Smart Cities, pp. 311–330. Springer, Cham (2018)
10. Sijini, A.C., Fantin, E., Ranjit, L.P.: Switched reluctance motor for hybrid electric vehicle. Middle-East J. Sci. Res. **24**(3), 734–739 (2016)
11. Raj, E.F.I., Balaji, M.: Analysis and classification of faults in switched reluctance motors using deep learning neural networks. Arab. J. Sci. Eng. **46**(2), 1313–1332 (2021)
12. Raj, E.F.I., Kamaraj, V.: Neural network based control for switched reluctance motor drive. In: 2013 IEEE International Conference on Emerging Trends in Computing, Communication and Nanotechnology (ICECCN), pp. 678–682. IEEE (2013)
13. Ferrer, J.R.: Barcelona's smart city vision: an opportunity for transformation. Field Actions Sci. Rep. J. Field Actions (Special Issue 16), 70–75 (2017)
14. Moik, B., Bobek, V., Horvat, T.: India's national smart city mission: analysis of project dimensions including sources of funding. Mednarodno inovativno poslovanje = J. Innov. Bus. Manag. **13**(1), 50–59 (2021)
15. Raj, E.F.I., Appadurai, M.: The hybrid electric vehicle (HEV)—An overview. Emerging Solutions for e-Mobility and Smart Grids, pp. 25–36 (2021)
16. Saharan, S., Kumar, N., Bawa, S.: An efficient smart parking pricing system for smart city environment: a machine-learning based approach. Futur. Gener. Comput. Syst. **106**, 622–640 (2020)
17. Sharma, H., Talyan, S.: IoT based smart car parking system for smart cities. In: Recent Trends in Communication and Electronics, pp. 372–374. CRC Press (2021)
18. Jan, B., Farman, H., Khan, M., Talha, M., Din, I.U.: Designing a smart transportation system: an internet of things and big data approach. IEEE Wirel. Commun. **26**(4), 73–79 (2019)
19. Aamir, M., Masroor, S., Ali, Z.A., Ting, B.T.: Sustainable framework for smart transportation system: a case study of karachi. Wirel. Pers. Commun. **106**(1), 27–40 (2019)
20. Aujla, G.S., Jindal, A., Kumar, N.: EVaaS: electric vehicle-as-a-service for energy trading in SDN-enabled smart transportation system. Comput. Netw. **143**, 247–262 (2018)
21. Schinkel, W., van der Sande, T., Nijmeijer, H.: State estimation for cooperative lateral vehicle following using vehicle-to-vehicle communication. Electronics **10**(6), 651 (2021)
22. Agarwal, V., Sharma, S., Agarwal, P.: IoT based smart transport management and vehicle-to-vehicle communication system. In: Computer Networks, Big Data and IoT, pp. 709–716. Springer, Singapore (2021)
23. Tientrakool, P., Ho, Y.C., Maxemchuk, N.F.: Highway capacity benefits from using vehicle-to-vehicle communication and sensors for collision avoidance. In: 2011 IEEE Vehicular Technology Conference (VTC Fall), pp. 1–5. IEEE (2011)
24. Wang, J., Liu, K., Xiao, K., Chen, C., Wu, W., Lee, V.C., Son, S.H.: Dynamic clustering and cooperative scheduling for vehicle-to-vehicle communication in bidirectional road scenarios. IEEE Trans. Intell. Transp. Syst. **19**(6), 1913–1924 (2017)
25. Luo, Y., Xiang, Y., Cao, K., Li, K.: A dynamic automated lane change maneuver based on vehicle-to-vehicle communication. Transp. Res. Part C Emerg. Technol. **62**, 87–102 (2016)
26. He, C.R., Orosz, G.: Saving fuel using wireless vehicle-to-vehicle communication. In: 2017 American Control Conference (ACC), pp. 4946–4951. IEEE (2017)
27. Poullikkas, A.: Sustainable options for electric vehicle technologies. Renew. Sustain. Energy Rev. **41**, 1277–1287 (2015)
28. Bawany, N.Z., Shamsi, J.A.: Smart city architecture: vision and challenges. Int. J. Adv. Comput. Sci. Appl. **6**(11) (2015)
29. Gohar, A., Nencioni, G.: The role of 5G technologies in a smart city: the case for intelligent transportation system. Sustainability **13**(9), 5188 (2021)
30. Bergenhem, C., Hedin, E., Skarin, D.: Vehicle-to-vehicle communication for a platooning system. Procedia Soc. Behav. Sci. **48**, 1222–1233 (2012)

31. Miller, J.: Vehicle-to-vehicle-to-infrastructure (V2V2I) intelligent transportation system architecture. In: 2008 IEEE intelligent vehicles symposium, pp. 715–720. IEEE (2008)
32. Mostowfi, S., Buttlar, W.G.: Vehicle-to-infrastructure and human-to-infrastructure models for smart civil infrastructure systems. In: International Conference on Applied Human Factors and Ergonomics, pp. 147–155. Springer, Cham (2020)

FACEIFY: Intelligent System for Text to Image Translation

Shyamapada Mukherjee, Satya Sanjay Nath, Gobind Kumar Singh, and Sumanta Banerjee

1 Introduction

The modern deep learning-based models are leveraged to enrich the power of the smart devices such that they can perceive their environment, process the data and respond. The Text-to-Image translation models have been one of the most efficient ways to unleash the power of deep learning technology. Text-to-Image translation primarily focuses on producing realistic and consistent images from the given textual descriptions. i.e. translating hand-written text to image pixels. For example: 'The man has a bushy grey moustache and a double chin' or something like 'The dog is very lean and has very little hair on his body'. Recently, the translation of text to image has gained a lot of attention in the field of Deep Learning and Research community due to the challenges this topic has. The current existing approaches do not bring the fine details and consistency with the descriptions provided (Fig. 1).

The challenge with text-to-image translation is actually solving two subproblems. First, to get the important details of the text, we need to know a text feature representation technique and second, use the features from the text to generate a consistent and detailed image. Both of these can be done as Deep Learning community has made enough advancement both in Natural Language Processing (NLP) [3] and Image synthesis. However, the challenge is with the multi-modality of the problem as conditioning image on textual descriptions is highly multi-modal.

This problem is, however, solved by the introduction of Generative Adversarial Networks (GANs) [4] by Ian J. Goodfellow in 2014 in which the Generator network and the Discriminator network works adversarially. The Generator network in GAN produces images conditioned on text and tries to escape the Discriminator network. The Discriminator's role is to distinguish between the images produced by the generator and the real image.

S. Mukherjee (✉) · S. S. Nath · G. K. Singh · S. Banerjee
National Institute of Technology Silchar, Silchar 788010, Assam, India
e-mail: shyama@cse.nits.ac.in

S. Mukherjee et al. (eds.), *Intelligent Systems for Social Good*,
Advanced Technologies and Societal Change,
https://doi.org/10.1007/978-981-19-0770-8_5

Fig. 1 Examples of generated images using textual descriptions

Our contribution in this work is at text-to-image translation focused on generating human facial images from the textual hand-written descriptions. The dataset that we used consists of 4076 distinct images. The human facial images have been randomly extracted from the CelebA dataset. The captions have been collected via crowdsourcing. The dataset was collected and provided to us by RiVaL group.

2 Related Work

Text-to-image is one of the most important research topics these days. The success in this area of generating images after the introduction of GAN has given the topic more research interest and has been revolving around GAN, especially after the success of Attention GAN [16], a network having generator and discriminator network stacked with attention at each level. However, this architecture of GAN also suffers many issues in training, modal collapsing and hyperparameter tuning.

Before the introduction of GANs, Variational Autoencoders (VAEs) [6] were popular for the generation of images but it had its limitations. Variational Autoencoder models rely heavily on the pixelwise reconstruction loss and produce pretty blurred images with lack of details and images that are inconsistent with the descriptions. Some current methods also incorporate VAE models with GAN model such as incorporating VAE model in the generator network of GAN. With the introduction of GAN and its better result in this field, the use of VAE in this area has gone obsolete.

In recent researches, many approaches using Attention GAN has received enough recognition by the addition of transformer model [13] and have also been benefited from the fact that these approaches prove to be more computational efficient. Recent researches have also explored the use of transformers for text-to-image translation.

GANs have left VAE and other approaches of image generation behind by producing much real and detailed images. But one of the problem GANs face is the instability in their training [15], and this instability in training the model makes it tough for generating higher resolution (e.g. 256 × 256) images. Many ideas were proposed to stabilise the training of GANs.

GANs as the base, conditional image generation [8] has also been studied. Several methods have been proposed to generate images from unstructured text. Reed et al. [11] proposed conditional PixelCNN for generating images using textual descriptions and object location. Nyugen et al. used an approximation Langevin sampling method [12] to generate images conditioned on text. Mansimov et al. used a model named AlignDRAW [7] to figure the alignment between the text and the generating canvas. Finally, Reed et al. using the idea of Conditional GAN successfully generated plausible 64 × 64 floral and birds' images [11]. Our paper deals with the generation of human faces conditioned on text.

Besides using a single GAN, a series of GANs have also been studied. Wang et al. [14] were successfully able to generate images of an indoor environment using S^2-GAN. Denton et al. [2] used a series of GAN with Laplacian Pyramid framework where at every level of pyramid an image was generated, which was conditioned on the image of previous stage and then added back to the input image to be useful for image generation in the next stage paving the path for Stack GAN [17].

3 Generative Adversarial Networks (GANs)

GANs consists of two networks (Fig. 2) working adversarilly and helping each other to achieve a common goal. The role of the generator is to generate images and the discriminator tries to distinguish between real training images and the images produced by the generator. The generator aims to fool the discriminator by producing such images that the discriminator becomes unable to distinguish between training image and the generated image. And thus after every iteration, the generator becomes better and produces more realistic images making it difficult for discriminator to distinguish. This is basically a minimax game. At the start of the training, the produced images are rejected by the Discriminator with high confidence and generator keeps on improving the produced images and the training stops at the time when the generator fools the discriminator by the produced images. The loss function for the network is given below:

$$min_G max_D L(D, G) = E_{x \sim p_r(x)}[log D(x)] + E_{z \sim p_z(z)}[log(1 - D(G(z)))]$$

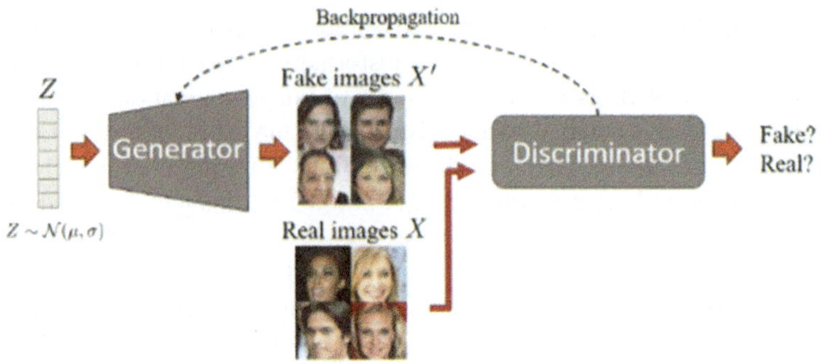

Fig. 2 Generative adversarial network

3.1 Discriminator

The discriminator figures out how to recognise the generator's phony information from genuine information. The discriminator punishes the generator for delivering doubtful outcomes. At the beginning, the generator delivers clearly counterfeit information, and the discriminator rapidly figures out how to tell that it's phony. The discriminator's training data come from two sources:

1. Genuine information occurrences, for example, genuine pictures of individuals. The discriminator utilises these cases as positive models during preparing.
2. Counterfeit information occasions made by the generator. The discriminator utilises these occurrences as negative models during preparing.

3.2 Generator

The generator part of the GAN is to how to make counterfeit information by considering the criticism from the discriminator. It tries to figure out how to make the discriminator label its produced information as genuine.

Generator preparing requires more tightly coordination between the generator and the discriminator than discriminator preparing requires. The part of the GAN that prepares the generator incorporates:

1. irregular information
2. generator arrange, which changes the arbitrary contribution to an information occurrence
3. discriminator arrange, which orders the created information
4. discriminator yield
5. generator misfortune, which punishes the generator for neglecting to trick the discriminator.

4 Dataset

4.1 Dataset Description

The dataset consists of 4076 distinct images. The human facial images have been randomly extracted from the CelebA dataset. The captions have been collected via crowdsourcing. The dataset was collected and provided to us by the RiVaL group (Figs. 3 and 4).

4.2 Prepossessing

The text was cleaned by removing stop words, correcting spelling mistakes and removing punctuation.

The images in the dataset contain human faces in different angles and perspectives and along with other the faces contained in the images, we need to isolate the faces from unnecessary details. We need MTCNN [18] model to extract the facial regions from the photos (Fig. 5).

Fig. 3 A man with medium-length grey hair and a receding hairline. His eyes are very small and they can barely be seen. He has got wrinkles on his face. He has got a grey beard with white patches

Fig. 4 A young woman with orange hair and big blue eyes. There is dark make-up around her eyes and her lips are thin. Her upper teeth are visible as she is smiling

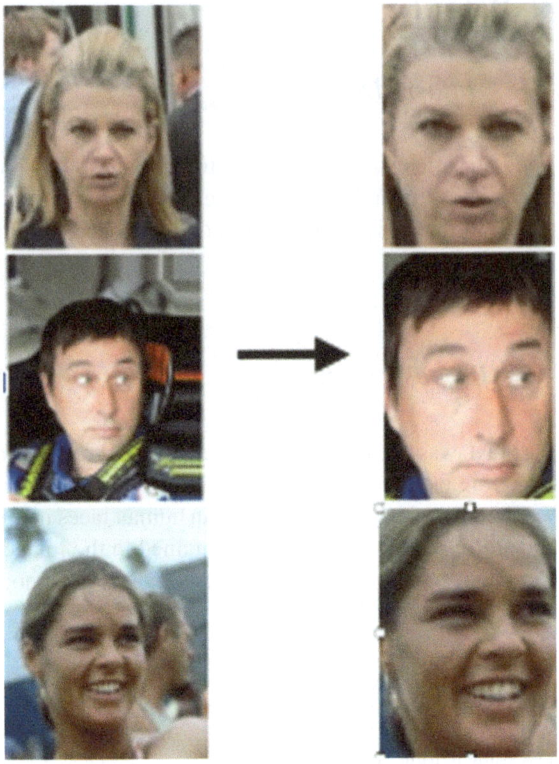

Fig. 5 Extracting facial regions from photos

5 Implementation and Architecture

5.1 Input to the Network

1. **Text Input**: We give the textual descriptions as input to the network so as to condition the GAN to produce images based on the descriptions and not random images from latent space. The generator takes as input a random noise vector concatenated with the text embedding from the GloVe [9] model. The generator outputs a 32×32 resolution image of the human face. The discriminator is given the training images and descriptions along with the images generated by the generator.
2. **Image Input**: The input images to the network are extracted from dataset and pre-processed using MTCNN [18] and are encoded using 2D Convolutional Layers.

The image input is passed through a series of convolution layers to extract the features. The textual input is passed through an embedding layer followed by a LSTM [5] layer. Both the encoded image and text inputs are concatenated into a single vector, which is passed through a dense layer to output a single number that is the probability of the image being real. The activation function used in the output layer is sigmoid.

5.2 *Model Architecture*

Our network is built using a Deep Convolution GAN (DCGAN) [10] architecture where the discriminator and generator are deep convolution neural networks. The basic network architecture consists of two constituent networks: a discriminator and a generator. The generator is responsible for generating new images and the discriminator distinguishes between the real and fake images. The network is trained using an adversarial minimax [1] approach where the discriminator is trained to maximise the probability of assigning correct label whereas the generator is trained to minimise the probability of assigning the correct label.

Furthermore, we give the textual descriptions as input to the network so as to condition the GAN to produce images based on the descriptions and not random images from latent space. The generator takes as input a random noise vector concatenated with the text embedding from the GloVe model. The generator outputs a 128 × 128 resolution image of the human face. The discriminator is given the training images and descriptions along with the images generated by the generator. The discriminator outputs a label to classify if the image generated by the generator is real or fake. In some cases, the images generated may be realistic but they may not match the text description. The discriminator needs to be trained to distinguish fake based on two parameters—unrealistic images as well as wrong mapping between text and images. The resulting network will be capable of generating realistic random human face images from text description (Fig. 6).

1. **Discriminator Network**: The baseline discriminator network proposed is based on the merge architecture. The image input is passed through a series of convolution layers to extract the features. The textual input is passed through an embedding layer followed by a LSTM layer. Both the encoded image and text inputs are concatenated into a single vector, which is passed through a dense layer to output a single number that is the probability of the image being real. The activation function used in the output layer is sigmoid.

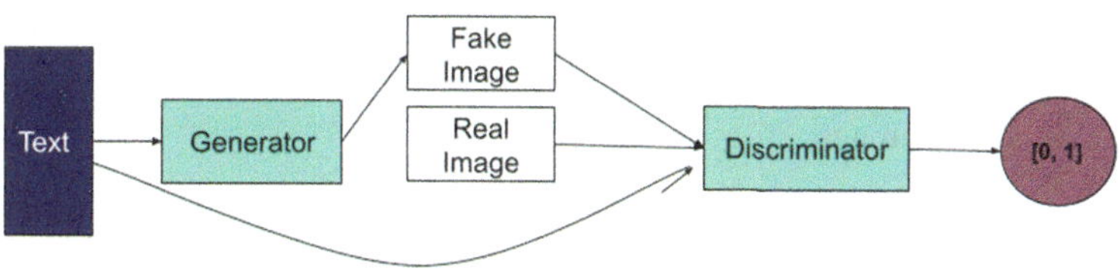

Fig. 6 Basic text-to-image GAN architecture

2. **Generator Network**: For the generator network, the text is similarly encoded using embedding and LSTM layer and given as input concatenated with a latent vector. The input is then passed through a series of transpose convolution layers to upsample the input into a 128 × 128 resolution image.

5.3 Training the Model

The training image size was 32 × 32 due to restricted resources. The discriminator network was trained with batch size 128, a base learning rate of 0.0005, and momentum 0.5. We trained the model for 200 epochs. The training was done using Google Colab platform.

6 Results

1. We have run the model for 200 epochs and plotted the loss and accuracy of discriminator and generator
2. The model obtained an inception score of 1.0001. State-of-the-art models for text-to-image synthesis have a score of 3.72 on the Oxford-102 flowers dataset.
3. As we see from the curve that the discriminator seems to be overpowering the generator and the model is thus collapsing. This is a case of convergence failure of GAN model. GANs are very unstable and challenging to train, and we are trying different techniques to stabilise the training process and converge the generator and discriminator and produce better results (Fig. 7).

7 Conclusion

Text-to-image synthesis is a relatively new area of research, which has seen fast growth since the introduction of GANs as they have proved to give much better results than other neural network models and autoencoders. The idea of incorporating such a technology into smart and resource-constrained devices would bring fruitful results in the future. The current approach has been able to generate human faces but have not been able to bring out the little details properly and the deals with the problem of lower resolution images (Fig. 8).

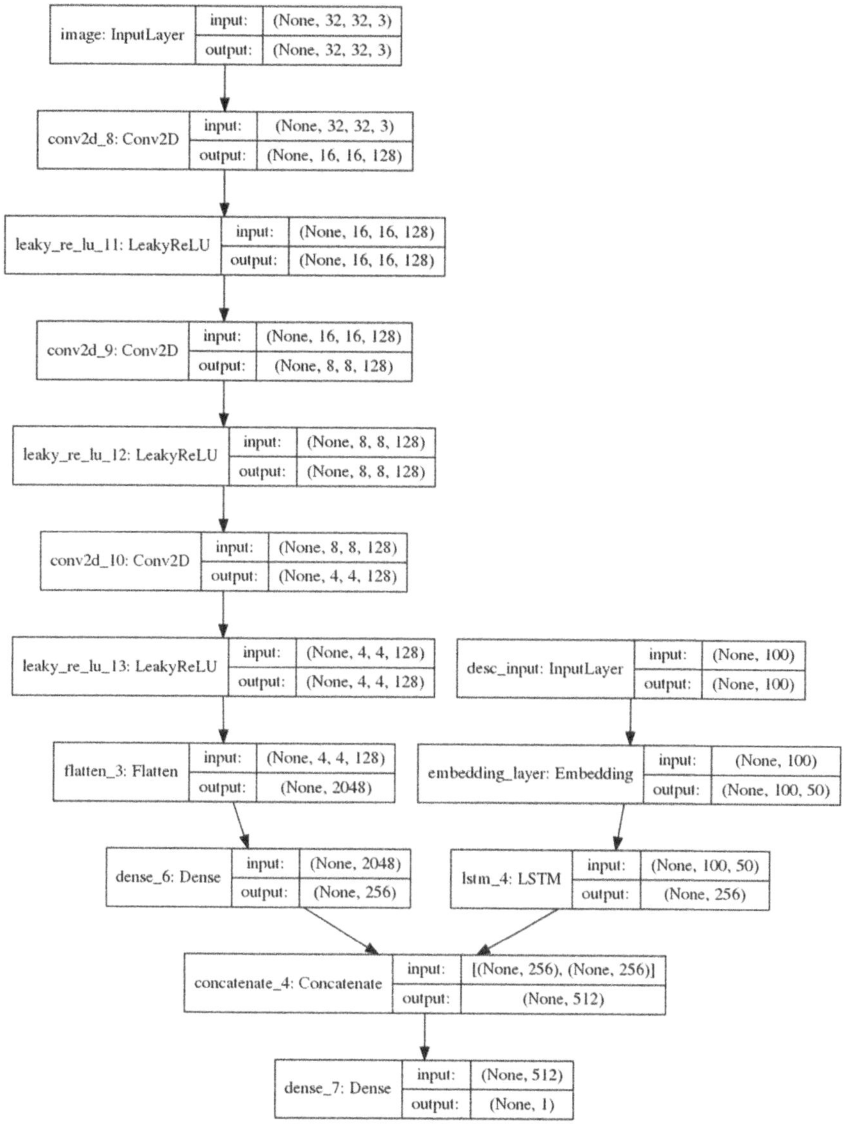

Fig. 7 Baseline discriminator network

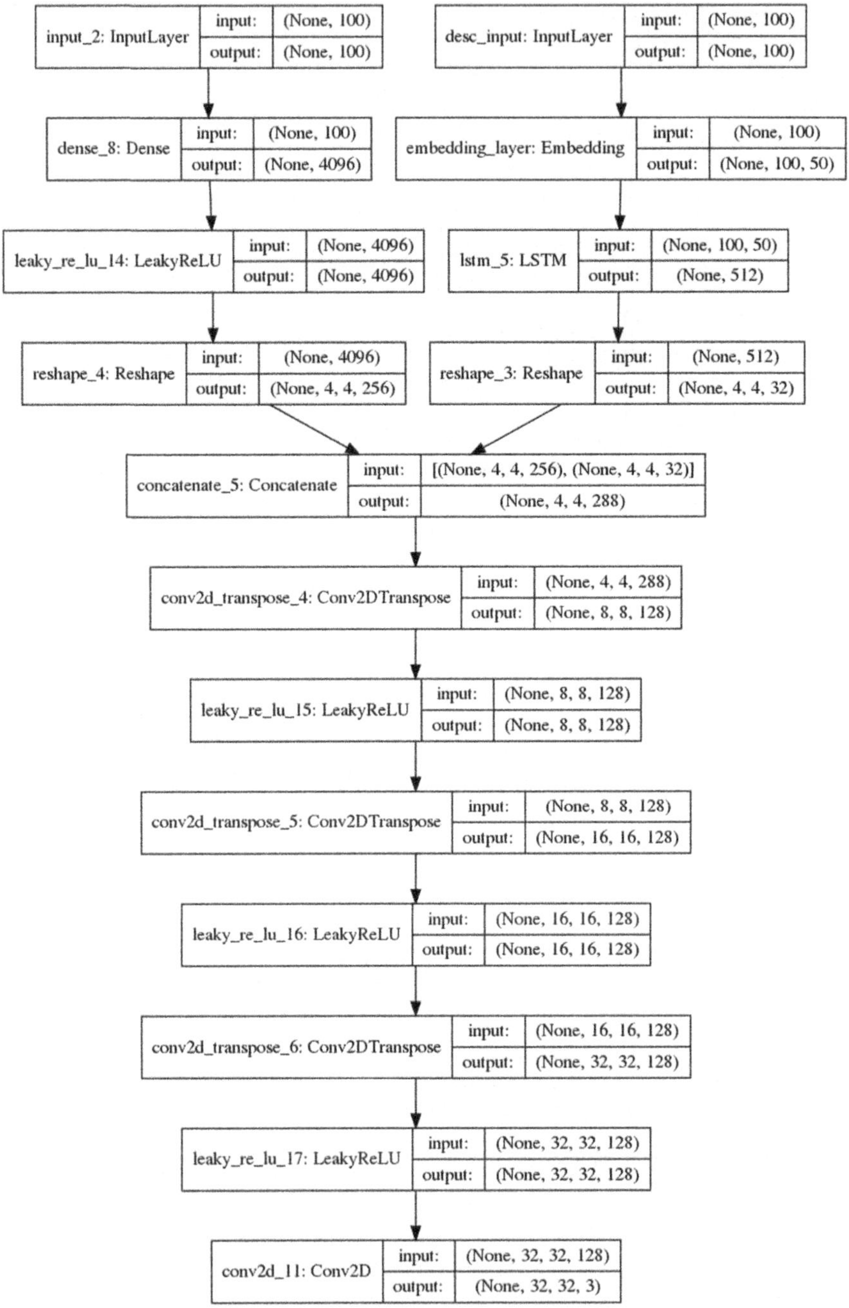

Fig. 8 Baseline generator network

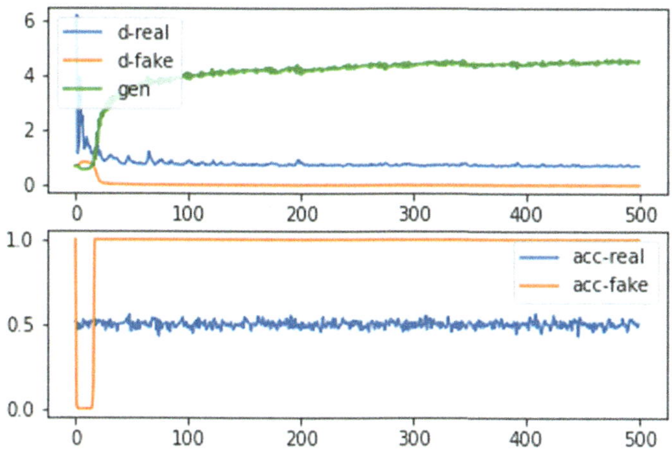

Fig. 9 Loss accuracy curve

8 Future Works

1. Stabilise the network to generate good quality images from text descriptions.
2. The model for generating images can also be used to generate videos, which can be achieved by generating each frame at a time and making a valid sequence out of all the images that connect together to give a logical meaning leads to a video.
3. This can further be extended to more industry use and complete automation of the police witness description and suspect detection system.
4. Can be used to clarify old and ancient black and white pictures using NoGAN training technique (Fig. 9).

References

1. Chi, P., Nau, D.S.: Predicting the performance of minimax and product in game-tree. CoRR abs/1304.3081 (2013). http://arxiv.org/abs/1304.3081
2. Denton, E.L., Chintala, S., Szlam, A., Fergus, R.: Deep generative image models using a Laplacian pyramid of adversarial networks. CoRR abs/1506.05751 (2015). http://arxiv.org/abs/1506.05751
3. Gelbukh, A.: Natural language processing. In: Fifth International Conference on Hybrid Intelligent Systems (HIS'05), p. 1 (2005)
4. Goodfellow, I., Pouget-Abadie, J., Mirza, M., Xu, B., Warde-Farley, D., Ozair, S., Courville, A., Bengio, Y.: Generative adversarial nets. In: Ghahramani, Z., Welling, M., Cortes, C., Lawrence, N.D., Weinberger, K.Q. (eds.) Advances in Neural Information Processing Systems, vol. 27, pp. 2672–2680. Curran Associates, Inc. (2014). http://papers.nips.cc/paper/5423-generative-adversarial-nets.pdf
5. Hochreiter, S., Schmidhuber, J.: Long short-term memory. Neural Comput. **9**(8), 1735–1780 (1997)

6. Kingma, D.P., Welling, M.: An introduction to variational autoencoders. CoRR abs/1906.02691 (2019). http://arxiv.org/abs/1906.02691
7. Mansimov, E., Parisotto, E., Ba, J., Salakhutdinov, R.: Generating images from captions with attention (2016)
8. Mirza, M., Osindero, S.: Conditional generative adversarial nets. CoRR abs/1411.1784 (2014). http://arxiv.org/abs/1411.1784
9. Pennington, J., Socher, R., Manning, C.: Glove: Global Vectors for Word Representation, pp. 1532–1543 (2014). https://doi.org/10.3115/v1/D14-1162
10. Radford, A., Metz, L., Chintala, S.: Unsupervised representation learning with deep convolutional generative adversarial networks (2015) (Comment: Under review as a conference paper at ICLR 2016). http://arxiv.org/abs/1511.06434. arxiv:1511.06434
11. Reed, S.E., Akata, Z., Yan, X., Logeswaran, L., Schiele, B., Lee, H.: Generative adversarial text to image synthesis. CoRR abs/1605.05396 (2016). http://arxiv.org/abs/1605.05396
12. Stuart, A., Voss, J., Wiberg, P.: Conditional path sampling of SDEs and the Langevin MCMC method. Commun. Math. Sci. **2** (2004). https://doi.org/10.4310/CMS.2004.v2.n4.a7
13. Vaswani, A., Shazeer, N., Parmar, N., Uszkoreit, J., Jones, L., Gomez, A.N., Kaiser, L.u., Polosukhin, I.: Attention is all you need. In: Guyon, I., Luxburg, U.V., Bengio, S., Wallach, H., Fergus, R., Vishwanathan, S., Garnett, R. (eds.) Advances in Neural Information Processing Systems, vol. 30, pp. 5998–6008. Curran Associates, Inc. (2017). http://papers.nips.cc/paper/7181-attention-is-all-you-need.pdf
14. Wang, X., Gupta, A.: Generative image modeling using style and structure adversarial networks. CoRR **abs/1603.05631** (2016). http://arxiv.org/abs/1603.05631
15. Xu, K., Li, C., Wei, H., Zhu, J., Zhang, B.: Understanding and stabilizing gans' training dynamics with control theory (2020). https://openreview.net/forum?id=BJe7h34YDS
16. Zhang, H., Goodfellow, I., Metaxas, D., Odena, A.: Self-attention generative adversarial networks. In: Chaudhuri, K., Salakhutdinov, R. (eds.) Proceedings of the 36th International Conference on Machine Learning, Proceedings of Machine Learning Research, vol. 97, pp. 7354–7363. PMLR, Long Beach, California, USA (2019). http://proceedings.mlr.press/v97/zhang19d.html
17. Zhang, H., Xu, T., Li, H., Zhang, S., Huang, X., Wang, X., Metaxas, D.N.: Stackgan: text to photo-realistic image synthesis with stacked generative adversarial networks. CoRR abs/1612.03242 (2016). http://arxiv.org/abs/1612.03242
18. Zhang, K., Zhang, Z., Li, Z., Qiao, Y.: Joint face detection and alignment using multi-task cascaded convolutional networks. CoRR abs/1604.02878 (2016). http://arxiv.org/abs/1604.02878

LiFi Technology: A Breakthrough for Massive Data Rates in Indoor Applications

Rahul Sharma, Devendra Singh Gurjar, Elvis Rahman, Aditya Raghav, Pranjal Shukla, and Vivek Mishra

1 Introduction

The data traffic is increasing at a tremendous pace, and along with it, the number of devices connected to the Internet is growing as well. Since the radio-frequency (RF) spectrum is already crowded, it would not be able to satisfy the rising demands [1, 2]. Light fidelity (LiFi) has emerged as a novel technology that uses visible light for downlink communication and lumination. Compared to RF systems, LiFi can provide a much higher signal-to-noise ratio (SNR). Moreover, the available unregulated bandwidth for LiFi communication is several times higher than the entire RF spectrum. It can also provide enhanced security as light does not penetrate through opaque objects.

In a broad sense, LiFi is a fast and cheap optical version of WiFi based on visible light communication (VLC). Typically, VLC uses light emitting diodes (LEDs) to transmit data wirelessly using intensity modulation and a photodiode at the receiver terminal [3]. Since LEDs are already prevalent at homes, offices, and streetlights, LiFi

R. Sharma · D. S. Gurjar (✉) · E. Rahman · A. Raghav · P. Shukla · V. Mishra
National Institute of Technology Silchar, Silchar, Assam, India
e-mail: devendra.gurjar@ieee.org

R. Sharma
e-mail: sharma_ug@ece.nits.ac.in

E. Rahman
e-mail: elvis_ug@ece.nits.ac.in

A. Raghav
e-mail: adityaraghav_ug@ece.nits.ac.in

P. Shukla
e-mail: pranjal_ug@ece.nits.ac.in

V. Mishra
e-mail: vivek_ug@ece.nits.ac.in

S. Mukherjee et al. (eds.), *Intelligent Systems for Social Good*,
Advanced Technologies and Societal Change,
https://doi.org/10.1007/978-981-19-0770-8_6

can use existing lighting infrastructure. Also, LiFi uses light for realizing communications. Therefore, it can be effectively used at places such as chemical manufacturing plants and aircraft cabins where RF transmission is too dangerous. Although LiFi is still in the developmental phase and it is not available for public use, several companies are working towards developing and marketing the LiFi based networking solutions.

1.1 Related Works

LiFi can very well resolve the issue of limited bandwidth for wireless communication for indoor applications. Several works have highlighted the importance of considering LiFi for such applications along with WiFi. In particular, the authors in [4] have appropriately classified LiFi from VLC and discussed the similarities and differences between the two. It has been shown that a unique LiFi-WiFi hybrid network can eliminate interference in LiFi and, at the same time, it can cover blind spots. But, there are chances of frequent handovers for mobile users, which can be resolved using fuzzy logic-based dynamic handover techniques. Further, authors in [5] have categorized existing network topologies related to LiFi and illustrated that the overall network performance can be significantly improved by implementing a hybrid LiFi/WiFi network. LiFi provides significant economic opportunities and will enable various industries to grow manifolds. However, there exist multiple scientific challenges. Load balancing in hybrid LiFi-WiFi networks is one such critical challenge. Hybrid LiFi-WiFi networks (HLWNets) can achieve higher throughput than LiFi or WiFi networks alone. Still, access point selection for HLWNets is tricky, and user mobility is of great concern [6]. However, this problem can be addressed using the mobility-aware load balancing (MALB) technique. In [6], the authors have proposed two novel methods considering single and multiple transmissions modes. In the single transmission method, users can be transferred between different access points. It has been shown that MALB with single transmission can suppress vertical handover at the cost of an increased rate of horizontal handover. In the multiple-transmission method, users can access both WiFi and LiFi, and vertical handover can be avoided entirely.

Moreover, the authors in [7] have focused on access point allocation and user mobility in HLWNets by suggesting a new load balancing scheme for HLWNets. Herein, the users are assigned to a type of network out of the three types of available networks in HLWNets, LiFi, WiFi, and LiFi-WiFi. Nodes in the primary two classes are handed over in the same network when needed. Both LiFi and WiFi can serve the users pertaining to the third category in case of light blockage.

The research work in [8] has investigated handover in HLWNets and introduced a novel handover method that is based on reference signal received power and not on user trajectory. The suggested handover system makes use of dynamic network performance that adapts to the node's velocity. Nodes having higher mobility are likely to be served by WiFi, whereas the LiFi access point supplies slow-moving

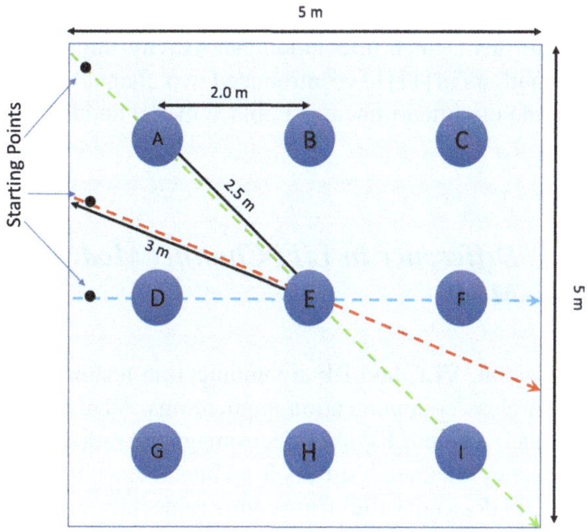

Fig. 1 Movement trajectories considered in [10] with luminaries colored blue

nodes with the greatest SNR [9]. Because the considered metric is widely utilized in existing handover methods, the approach does not need any new signaling between the user and the access point.

1.2 Channel Model for LiFi

Although, LiFi has been extensively studied, the existing research literature for channel modeling and characterization is somewhat limited. In reality, each LiFi transceiver design must include the channel model. For designing realistic LiFi transceivers, realistic channel models will help categorize the achievable performance.

(i) Most of the works on VLC channel modeling consider the transmitter and receiver to be fixed, which is not very realistic. Practically, the receiver can be mobile. Also, it has been characterized by the assumptions of vacant rooms neglecting the humans' interference or any other objects. In [10], a more realistic approach has been used to come up with a mobile VLC channel model using non-sequential ray tracing. This model visualizes a standard room having multiple luminaires, objects, and human interference. For this model, a person having an appropriate device steps in through various trajectories inside the room, as shown in Fig. 1. The channel impulse response and path loss are then obtained for these trajectories. It has been observed that the received power changes considerably with the change of location inside the room.

(ii) Most of the channel models, including given in [10], consider that the receiver always kept vertically facing directly upward. Such an assumption is not suitable for all devices as the majority of users hold devices in a position comfortable to them. Such orientation can heavily impact the user's throughput. Recently, authors in [11] have introduced two-channel models for stationary users, namely, the modified truncated Laplace model and the modified Beta model.

1.3 Difference in LiFi Channel Mode and WiFi Channel Mode

In general, VLC and RF communication technologies are utilized in a wide range of wireless communication applications. VLC employs light as a communication channel, whereas RF uses electromagnetic radiation. Because WiFi employs a radio frequency channel, it suffers from interference from adjacent access points (routers), but LiFi does not suffer from comparable interference concerns as it employs an LED channel. The LED channel outperforms wireless RF systems like WiFi in terms of increased security, electromagnetic interference reduction, and ultra-dense cellular reuse. As a result, LiFi might provide considerable relief from overcrowding in the RF spectrum and meet future connectivity needs as required for the Internet of Things (IoT). A basic block diagram consisting of all the necessary components required for realizing LiFi communications is depicted in Fig. 2. There are plenty of applications of LiFi due to its offered attributes. Some specific advantages can be noted as follows.

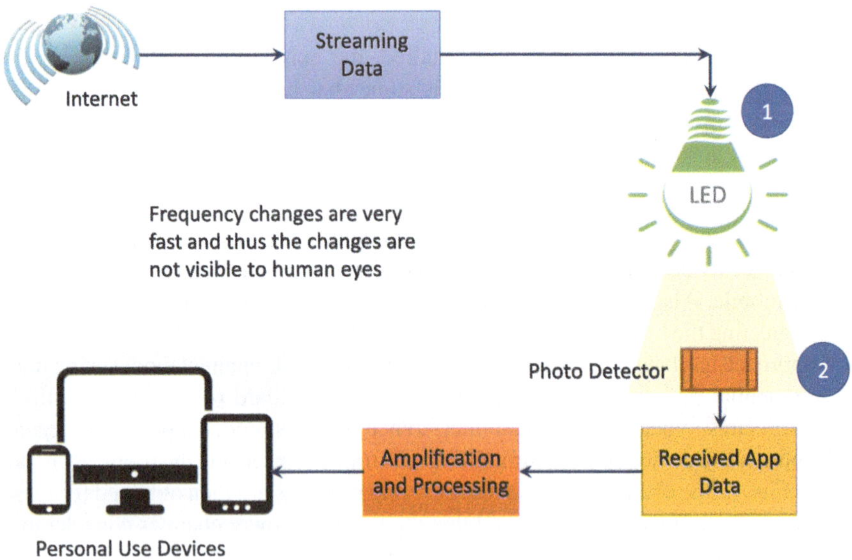

Fig. 2 Basic blocks for LiFi communications

1. It has higher bandwidth and so overcomes the RF transmission capacity constraint.
2. When both source and receiver are in line of sight (LoS) within the same room, VLC communication works effectively. Nobody in another room can intercept data communication based on VLC. As a result, unlike RF transmission, VLC provides better security.
3. The VLC source is used for both lighting and communication, and it consumes very little power that makes VLC a power-saving system.
4. It is unaffected by EM radiation from RF transmissions.

The main concern with LED channels using VLC communication is that other ambient light sources might cause interference with VLC-based communication. Moreover, it has a short range of coverage. Integrating VLC with a wireless setup is complicated. Other issues with the VLC system include air absorption, shadowing, and beam dispersion, among others. Both the source and the receiver must be in the LoS. As a result, achieving non-LOS communication is challenging. The data transfer speed is greater in the LED channel than the RF channel, and the security risk is also less using LED media than the RF channel. Also, the higher frequencies are attenuated more strongly in LEDs, which causes another problem. This necessitates an iterative trade-off between boosting transmit power at high frequencies and tolerating a smaller total throughput from expanding the constellation size at lower frequencies. To increase the performance of LED communication channels, sophisticated modeling of the physics of light production in semiconductor junctions can be used. Indeed, the nonlinear dynamics of photon emission events provide a rich source of inspiration for developing LED communication models. Novel DSP approaches for solving challenges like nonlinear distortion with minimal complexity and latency are also inspired by a trustworthy LED channel model. The LED channel models and DSP approaches encompass current major research and development trends in LiFi with the grand goal of enabling next-generation communication and IoT applications. The LED channel enables secure communication since it uses LOS communication within the room, but the RF channel are vulnerable for eavesdropping because RF signals can pass through walls and be intercepted by someone in another room.

LiFi employs ordinary LEDs to allow data to transmit at speeds of up to 224 gigabits per second. Light can be used to transmit data with this technology because light intensity changes faster than the sensitivity of human eyes. LiFi's transmission range is 100 times greater than WiFi's. This technology's data transfer may be accomplished using light. The efficient LEDs are the most important components of this system. LEDs' ON/OFF activity allows for data transfer in the form of binary codes, but the human eye is unable to perceive this transformation. Thus the bulbs appear to have a consistent intensity. As WiFi communication uses an RF channel for transmission, radio signals, antennas, and routers become the three basic parts of a wireless network. WiFi networking is only feasible because of radio waves. WiFi cards are installed on PCs and cell phones. Antenna and routers send the radio signals, and WiFi receivers, such as computers and mobile phones with WiFi cards, pick up the signals.

2 Networking in LiFi

The broad spectrum of visible light is utilized for downlink transmissions in LiFi, whereas the infrared-based communication in the uplink. LiFi refers to broader network systems that incorporate multi-user, bidirectional, and broadcast communication. Multiple access points (APs) can be used in a LiFi network. These numerous APs form microcellular systems to render high-data-oriented connectivity to various wireless users simultaneously. Two-way communication can be formed among an AP and user equipment (UE) in a LiFi network, allowing an AP to serve several users at once. Further, backhaul links connecting APs and the gateway are necessary to connect globally. As a sequence, powerline communications, power-over-Ethernet, or optical fiber can be used to offer these backhaul connections. Moreover, existing LED lighting installations can be used for the LiFi downlink.

Infrared spectrum has been preferred for uplink communications because visible light considered for the downlink may cause distractions for the mobile user. In addition, it has the benefit of preventing interference between the uplink and downlink, allowing for simultaneous transmission. Despite some preliminary investigations on the infrared-based uplink, more extensive research is needed for such systems. A variety of RF-based communication technologies, such as Bluetooth, WiFi, can also be explored. Although these technologies are widely available, they may cause interference with existing RF wireless networks. However, only RF/VLC hybrid connection employing VLC for downlink can offload significant data traffic while maintaining low latency.

Frequent handover due to mobility of users, multiple access considering a large number of users, and co-channel interference are essential things to handle in a complete LiFi network, as shown in Fig. 3. In LiFi, users can face horizontal and vertical forms of handovers. The authors in [12] have conducted an early investigation

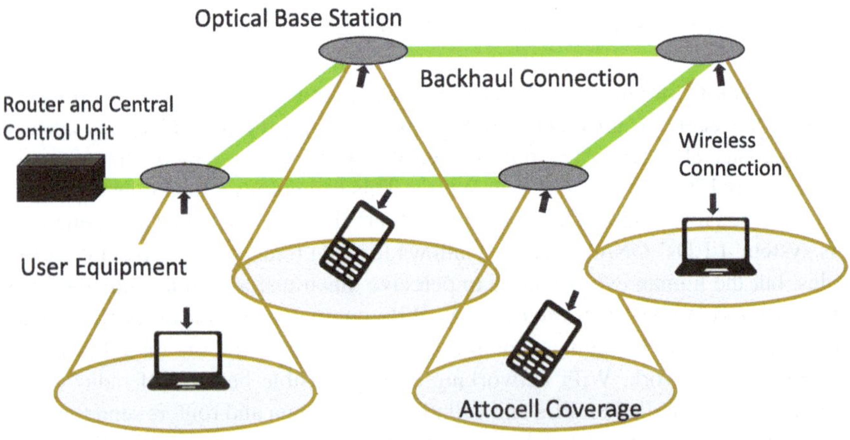

Fig. 3 The illustration of LiFi networks applied to indoor wireless networking

on the horizontal handover method in LiFi networks. A shift in the serving AP from one radio access technology (RAT) to another is referred to as a vertical handover. Mobile users may be moved from a LiFi AP to a WiFi AP if none of the LiFi APs can provide a stable link. This can happen if the user's speed is high or the dwell time in a cell is too short of establishing a meaningful communication link. When a user slows down and comes within range of a lightly loaded LiFi AP, it may be beneficial in handover to that LiFi AP to alleviate the WiFi network and run more effectively (e.g., ensuring fewer packet collisions). A vertical handover system based on the forecast of uncertainty can reduce transmission times considerably. In [13], the authors have suggested a vertical handover approach considering the estimates of uncertainty metrics.

3 Challenges in Deployment and Operations

Energy and big data management in smart cities bring together diverse professional viewpoints on essential subjects that underpin successful smart city development. The majority of the worries regarding 5G's alleged detrimental impact on health arise from the fact that its cell towers are so different in construction from those supporting today's 3G and 4G cellular networks. As a result, LiFi is the most critical choice for achieving high data rates and meeting smart city requirements. LiFi can handle a large number of users who are required to connect everything to the internet, often known as the IoT. Because of their rapid reaction time, extended operating lifetime, and inexpensive cost, LEDs are the best choices for LiFi system deployment. It does, however, impose certain restrictions. The following section discusses an overview of the difficulties and obstacles of deploying a LiFi system.

3.1 Techniques Involved in Transmission and Modulation

Commercial LiFi communication currently has a maximum operating distance of 1-50 m. However, because LiFi is a LoS communication system, this distance is much shorter than its radio frequency equivalent. Furthermore, the data speeds that may be achieved with LiFi are restricted to some lower amount. However, a substantial investigation is underway in this area to increase data rates by experimenting with various techniques and different types of materials of LED and photo-sensors. Even in LiFi, high-speed communication may be achieved by adopting a modulation method with high spectral efficiency. The use of trichromatic LEDs instead of phosphorescent LEDs resulted in a three-fold increase in throughput.

3.2 Optical Wireless Channel for Indoor Communications

The LiFi based communication is highly affected by the presence of LoS transmission, and any misalignment can significantly degrade performance. However, utilizing non-LoS communication channels, the shadowing effect created by obstructing the direct ray route among the transmitting LED and photo-sensor may be exploited to reconstruct data.

3.3 Receiver Device and Its Properties

In LiFi systems, photosensors at the receiver side detect the variation in light intensity. Image sensors or photodiodes might be used as photosensors. The photodiode transforms the amount of light detected into a photo-current. The performance of different photodiodes has been well investigated. It has been observed that avalanche photodiodes can be utilized for LiFi communication because of their tinier size, lower cost, and quicker reaction time. Nevertheless, the LiFi system's performance may be in the presence of sunlight and other bright fixtures. Thus, interference impacts can be considerably minimized by employing specialized filters and selecting a suitable receiver configuration. The LiFi communication system discovered that photodiodes for fixed receivers and image sensors for mobile users are ideally matched.

3.4 MIMO Optical Wireless Communications

Owing to the configuration limitations of small beamwidth receivers, using standard RF multiple-input multiple-output (MIMO) methods for LiFi poses a significant predicament. Even minor misalignment affects the communication quality. As a result, developing effective receivers and sensible transmitter deployment-based modulation techniques to enable MIMO is a promising area to explore for further investigation.

3.5 Cross-Layer Load Balancing

In case of heavy traffic or blockage, the network traffic must be transferred from LiFi to RF or vice versa. In order to perform this task, appropriate load balancing must be implemented. The parameters of the physical layer and media access control (MAC) layer can be optimized at a central unit load balancing system by utilizing a two-tier buffer structure. In terms of data rates, dynamic cross-layer load balancing results in overall increased performance.

3.6 Illumination Requirements

A good LiFi system should be able to handle varied dimming levels. Pulse width modulation and continuous current reduction are two methods that can perform dimming operations in indoor LiFi communication systems.

3.7 User Movement Modeling

User mobility and device orientation should be studied and simulated while constructing a LiFi network to research and offer smooth connections. The random waypoint model is often utilized to mimic a node's movement. Nevertheless, in real-world settings, it is unworkable. Users at a retail mall will undoubtedly move differently than those in an office setting. Furthermore, the user data rate would be affected by the unpredictable orientations of LiFi receivers, which should be adequately simulated. An orientation-based random waypoint mobility design was developed mainly for LiFi mobile devices, and it was evaluated considering the rate of handover in actual circumstances.

4 Coexistence of LiFi and WiFi

High signal strength in the interior environment of a building is rare, except in dense WiFi networks where contention is possible. Similarly, ultra high speed LiFi communications can be realized only when the transmitter and receiver are in LoS. On the other hand WiFi can be used within its range even if the transceivers are not in LoS. This motivates the idea of integrating WiFi with LiFi for more stable and faster communications as depicted in Fig. 4. Conventionally, the WiFi signal's intensity weakens as the distance between AP and users grows, even with a strong WiFi connection in a structure having various sorts of walls and other obstacles. As a result, if the signal strength in one room is significantly reduced, WiFi users will have poor connectivity and sluggish speeds. High interference signals from adjacent WiFi APs and lack of bandwidth can also cause slow connectivity when a WiFi AP's limited bandwidth is shared by numerous active users.

To achieve multi-Gbps peak data speeds, the WiFi development explores higher frequencies with the new spectrum for indoor scenarios (at 60 GHz) so that it can serve numerous users concurrently. At the same time, tri-band (2.4, 5, and 60 GHz) IEEE 802.11ad wireless local area network (WLAN) deployments are starting to enter the customer business. On the other hand, optical wireless communications systems explicitly based on VLC technology, also known as LiFi, offer dual-functionality to transmit data based on the intensity of optical sources. WiFi has been improved as the primary scalability element for wireless capacity. However, notably, WiFi's long-

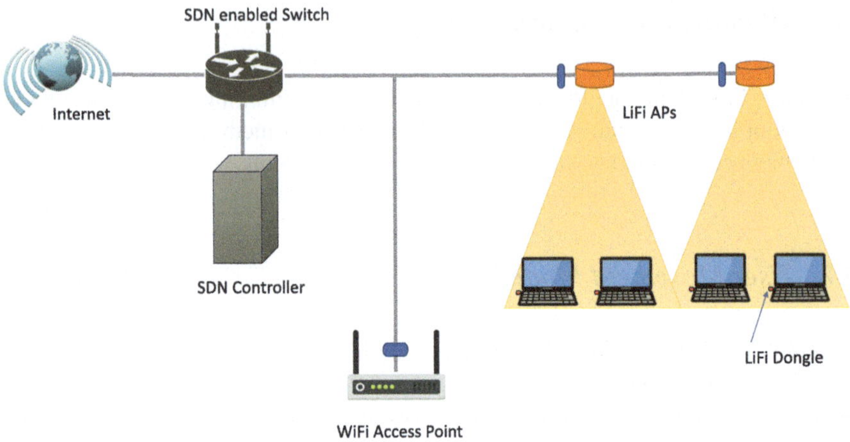

Fig. 4 Hybrid WiFi-LiFi network

term performance might be harmed in dense deployments as the carrier's capacity decreases.

For instance, the first node who discovers an unused channel is permitted to initiate the communications, regardless of the channel's quality. However, if another user with a better channel requests service later, the request will not be fulfilled since the initial link will not be disrupted due to the CSMA/CA requirement that the next transmission will only begin if the channel is free. The problem is made worse because the increasing adoption of IP video streaming results in higher data usage and the necessity for continuous gap-free data transmission. As a result, WiFi uses concurrent multi-user transmission as the following step, comparable to Long-Term Evolution, allowing multiple users with multiple-input and multiple-output transmissions. In densely populated areas, cooperative behavior is essential. It is also worth considering beamforming between the neighboring APs. However, defining such a new form of simultaneous communication will need a significant standard effort. Furthermore, with a higher number of antennas, there are complexity constraints. A potentially intriguing study topic is the coexistence of WiFi and LiFi. Figure 5 shows that the signal is available in all the directions in a standalone WiFi network, but the data rate of each connected device is low, and the load on the WiFi access point is relatively high. On the other hand, the data rate for connected devices with LiFi standalone networks is very high. Still, if there is a blockage of LiFi light signals in between the device and the LiFi access point, the data rate drops to zero, even when the device is within the range of LiFi coverage as shown in Fig. 6. Further, Fig. 7 depicts that by combining both types of access points in a hybrid network, we can overcome the shortcomings of both types of networks. For example, if there is a blockage for LiFi light signals, the WiFi access point may connect that device. When LiFi connectivity is available, there is no need to waste WiFi bandwidth.

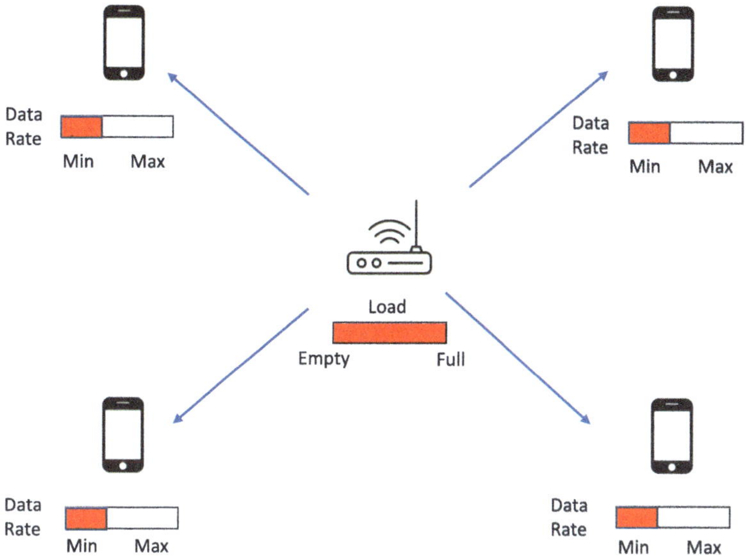

Fig. 5 WiFi standalone network

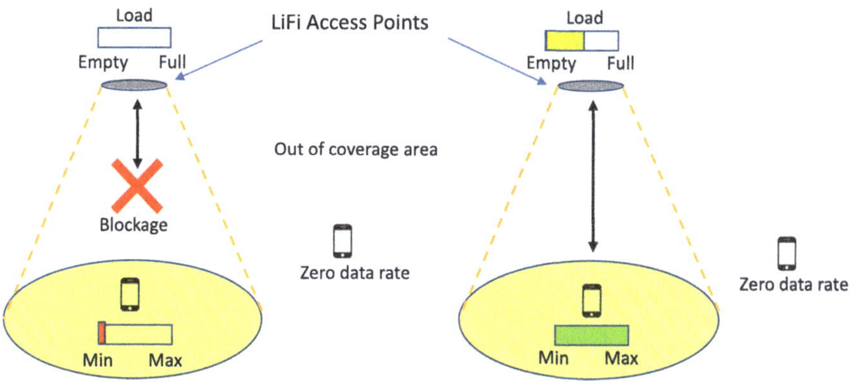

Fig. 6 LiFi standalone network

4.1 Load-Balancing Techniques for Hybrid LiFi-WiFi Networks

When multiple devices are running on the same WiFi, the connection speed can be significantly reduced as shown in Fig. 5. However, in a LiFi access point, one can access a high-speed connection in the presence of multiple devices connected through the same AP if they are in the vicinity of the AP. If the users are not in the vicinity of LiFi AP, they will not be served by that particular AP as shown in Fig. 6

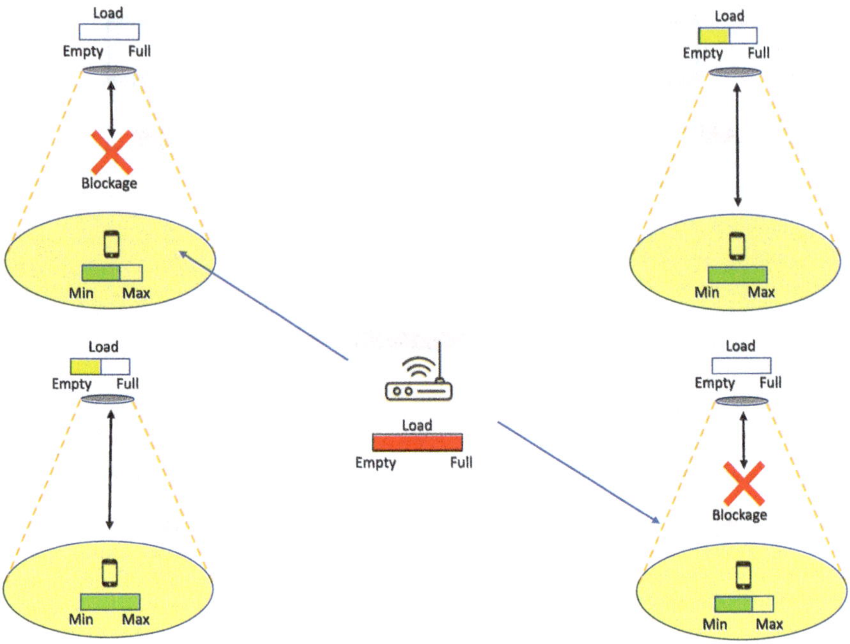

Fig. 7 WiFi-LiFi hybrid network

Therefore, the interaction between WiFi and LiFi networks will undoubtedly reduce the chance of link failure and improve system performance. This integration can be considered as cross-network technology where a lot of work needs to be done related to load balancing, channel allocation, user mobility, hybrid AP planning, etc. There are several ways to solve the load balancing problem:

- One can adopt a method involving nonlinear programming mixed-integers. However, this comes with high computational complexity.
- A game-theory-based categorized strategy has been developed that needs less computing complexity but provides only an asymptotic solution to the global optimum.

The later method turned out to be universal in solving complex problems. However, as a rule, this heuristic approach has low tractability, making it difficult to conduct an analytical test and prove its effectiveness. A recent study examined the load distribution of LiFi/WiFi hybrid networks in extreme cases, and it was found that with appropriate load balancing solutions for the user, service and overall quality improved by up to 80% compared to specific solutions. It should be noted that the service level refers to the level of quality of user service, which is defined as an average value.

In addition to improving the performance of the communication system for transmission and energy savings, it is also subject to evaluation. In a comprehensive study of the RF/ VLC hybrid network, the load balancing function is used in terms of energy consumption. LiFi and RF-based hybrid networks can significantly improve energy efficiency.

4.2 Framework for Parallel Transmission in LiFi

The parallel transmission in LiFi technology takes into account an internal wireless network with several LiFi APs. These APs are built inside LED ceiling lights that are perpendicularly pointing downwards [14]. They are organized in a square lattice topology to represent standard illumination arrangement in an indoor environment, such as an office. Interfering signals are regarded as noise by the APs, which reuse optical frequency bands. Each AP is made up of several LEDs with different wavelengths for lighting purposes (Table 1).

4.3 Mobility Modeling

A few studies have looked into user mobility models for indoor environments. These studies are typically based on a certain floor plan that includes rooms and corridors [14]. It is presumed in this article that there are no cubicles or corridors, and users are free to roam around. The random waypoint, which is a widely used synthetic model for mobility, is used to mimic freely moving individuals. Users travel in a zigzag pattern from one waypoint to the next, with the waypoints allocated at random. The user walks about in a wide outdoor area, such as a 1000 m by 1000 m zone, in the original RWP model, changing pace as it approaches each waypoint. The user's pace is kept constant for a brief length of time in an indoor situation where the distance between two destinations is quite short. An excursion refers to the user's mobility during this time. When the current trip is completed, the user selects a new pace and

Table 1 Parameter of different wavelengths [14]

Wavelength (nm)	450	500	600	650
Luminous efficacy	27	205	340	68
PD responsivity	0.15	0.23	0.37	0.44
Intensity percentage (%)	10	55	20	15
Optical output power	9.3	6.7	1.5	5.5
Modulated optical power	3.6	2.3	1.5	1.2

continues on their journey. The user's speed is considered to be evenly distributed between 0 and 2 v, with v indicating the average speed.

4.4 Light Path Blockage Model

When a LiFi link is obstructed, two variables influence its performance:

(i) how often the blockage happens
(ii) how long the obstruction lasts.

The first component impacts the average user throughput, whereas the second influences the handover rate caused by bottlenecks [14]. To characterize light-path obstructions, two metrics are used: the occurrence rate and the occupancy rate. The incidence rate, indicated as, is the average number of blockages that occur per unit of time. The Poisson point process is believed to govern this rate, which is commonly used to simulate random occurrences like packet arrival at a switch. The occupancy rate is the percentage of time during which the user encounters obstacles. We presume that various APs are blocked independently. Signals from several APs may be blocked in practice. This may be seen in the case of a high occupancy rate, where signals from numerous sources are combined. At the same time, APs are likely to be prohibited.

4.5 Choice of Subflows

A number of subflows are evaluated, which are chosen according to the signal strength strategy rule, which states that each user selects a number of APs that offer the greatest SINR values. A subflow can fail in one of two ways

- Loss of connection
- Increased traffic congestion.

In the first case, a handover can be used to allow a new subflow to join the existing link and maintain connectivity. There are two choices in the second case. One solution is to switch out the congested subflow for another. However, this might have an impact on other users' APs results, perhaps causing a chain reaction. The alternative is to avoid using the crowded subflow and redirect traffic to other subflows. Because the congestion conditions among APs are equivalent under the assumption of evenly dispersed users, this alternative is used in this study.

5 Current Standardization of LiFi

The International Telecommunication Union (ITU-T) has been involved in LiFi since 2015 as part of the G.vlc project. The ITU G.9991 recommendation for high-speed connectivity for VL and IR products was published in 2017. A high-quality network is crucial for the development of new technologies. Since 2015, the International Telecommunication Union (ITU-T) has been advising on high-speed connectivity for VL and IR products. The 802.15.7r1 Task Group (TG) was created in 2015 by the Institute of Electrical and Electronics Engineers (IEEE) to drought a revision to the IEEE 802.15.7 standard, which defines the PHY and MAC layers for short-range wireless optical communication utilizing VL [15]. The goal of 802.15.7r1 was to include IR and near-UV wavelengths, as well as alternatives like optical camera communications and LiFi, in addition to VL. The IEEE 802.11 working group for local area networks created the amendment TG-802.11bb in 2018, with the goal of developing technical requirements for enabling low-cost and low-energy devices for LiFi mass-market and assisting manufacturers and operators in providing universal 802.11 components. The aim is to reuse the 802.11 MAC for optical spectrum communications, which has a low adoption rate, and create a PHY that supports UL and DL operations in an optical wavelength range ranging from 380 to 5,000 nm. An association called LiFi Consortium was formed by a group of big telecommunication companies and industries in order to improve high speed wireless communications based on the optical wireless technology to provide considerable relief from overcrowding in the RF spectrum. The job of bringing high-speed, ubiquitous LiFi technology to market has begun at the University of Strathclyde in Scotland.

6 LiFi Application in Industries

LiFi offers a large variety of applications in live streaming, hospitals, workplaces, manufacturing facilities, schools, retail, and many more. LiFi can deliver communication 200 times faster than WiFi theoretically typically at around more than 200 Gbps. The benefits are summarized as follows:

1. GPS capabilities that are extremely accurate
2. Eco-friendly
3. Enhanced indoor connection
4. There are no health hazards
5. Very cost-effective
6. Faster than a typical network connection and there is no electromagnetic interference
7. Enhanced security.

LiFi has great potential to transform overall experience of passengers and in flight connectivity if used in aerospace. Transmitting data through light is attracting the interest of several airline companies. According to their analysis, using LiFi on airplanes would save the equivalent of 20 passengers' weight per flight. The reason being optical fibers weigh an order of magnitude less than copper. Airlines will be able to eliminate Internet connection equipment beneath seats, which will result in significant weight loss overall, increasing their profit per flight. However, owing to the installation of LiFi in the cabin, flight attendants will be able to approve in-flight payments instantly in the future. LiFi will help pilots and provide new applications for passengers since it is safer than WiFi and does not pose any risk of electromagnetic interference.

Major aerospace companies are looking into installing LiFi in their plane cockpits to connect the pilot's controls and equipment in a more accessible and safe manner. In the future, LiFi has the potential to improve the passenger experience. The inclusion of LiFi in the cockpit reduces the number of cables and reduces weight significantly. While WiFi has the potential to solve this problem, it must be used with caution due to its susceptibility to outside influence and hackers. On the other hand, LiFi is safer since, unlike WiFi, it cannot be transmitted through hulls and windows, making tapping from outside impossible. Similar to these improvements, other industries are benefiting from the development of LiFi technology.

7 Conclusion

This chapter discussed various aspects of LiFi, including the receiver and transmitter architecture and channel model. Also, we have emphasized the potential applications of LiFi in industries and smart cities. Various challenges related to user mobility, security, and load balancing have been presented. Further, we have motivated the concept of hybrid WiFi and LiFi to improve the reliability and achievable rate and end-users. Due to their distinct communication characteristics, both VLC and RF communications coexist. VLC, for example, is best for short-range communication, but RF is best for long-range communications. In particular, VLC offers suitable attributes for interior use, whereas RF can provide a better experience in indoor/outdoor applications.

References

1. Mandloi, M., Gurjar, D., Pattanayak, P., Nguyen, H.: 5G and Beyond Wireless Systems: PHY Layer Perspective. Springer Nature (2020)
2. Gurjar, D.S., Upadhyay, P.K.: Performance analysis of zero-forcing-based multiple-input multiple-output two-way relaying in overlay device-to-device communications. IET Commun. **10**(6), 699–708 (2016)
3. Pathak, P.H., Feng, X., Hu, P., Mohapatra, P.: Visible light communication, networking, and sensing: a survey, potential and challenges. IEEE Commun. Surv. Tutorials **17**(4), 2047–2077 (2015). https://doi.org/10.1109/COMST.2015.2476474

4. Haas, H., Yin, L., Wang, Y., Chen, C.: What is LiFi? J. Lightwave Technol. **34**(6), 1533–1544 (2016). https://doi.org/10.1109/JLT.2015.2510021
5. Haas, H., Yin, L., Chen, C., Videv, S., Parol, D., Poves, E., Alshaer, H., Islim, M.S.: Introduction to indoor networking concepts and challenges in LiFi. IEEE/OSA J. Opt. Commun. Netw. **12**(2), A190–A203 (2020). https://doi.org/10.1364/JOCN.12.00A190
6. Wu, X., Haas, H.: Mobility-aware load balancing for hybrid LiFi and WiFi networks. IEEE/OSA J. Opt. Commun. Netw. **11**(12), 588–597 (2019). https://doi.org/10.1364/JOCN.11.000588
7. Wu, X., Haas, H.: Load balancing for hybrid LiFi and WiFi networks: to tackle user mobility and light-path blockage. IEEE Trans. Commun. **68**(3), 1675–1683 (2020). https://doi.org/10.1109/TCOMM.2019.2962434
8. Wu, X., O'Brien, D.C., Deng, X., Linnartz, J.P.M.G.: Smart handover for hybrid LiFi and WiFi networks. IEEE Trans. Wireless Commun. **19**(12), 8211–8219 (2020). https://doi.org/10.1109/TWC.2020.3020160
9. Soltani, M.D., Arfaoui, M.A., Tavakkolnia, I., Ghrayeb, A., Safari, M., Assi, C.M., Hasna, M.O., Haas, H.: Bidirectional optical spatial modulation for mobile users: toward a practical design for lifi systems. IEEE J. Select. Areas Commun. **37**(9), 2069–2086 (2019). https://doi.org/10.1109/JSAC.2019.2929381
10. Miramirkhani, F., Narmanlioglu, O., Uysal, M., Panayirci, E.: A mobile channel model for VLC and application to adaptive system design. IEEE Commun. Lett. **21**(5), 1035–1038 (2017). https://doi.org/10.1109/LCOMM.2017.2651800
11. Arfaoui, M.A., Soltani, M.D., Tavakkolnia, I., Ghrayeb, A., Assi, C.M., Safari, M., Haas, H.: Measurements-based channel models for indoor LiFi systems. IEEE Trans. Wireless Commun. **20**(2), 827–842 (2021). https://doi.org/10.1109/TWC.2020.3028456
12. Vegni, A.M., Little, T.D.C.: Handover in VLC systems with cooperating mobile devices. In: 2012 International Conference on Computing, Networking and Communications (ICNC), pp 126–130 (2012). https://doi.org/10.1109/ICCNC.2012.6167395
13. Liang, S., Tian, H., Fan, B., Bai, R.: A novel vertical handover algorithm in a hybrid visible light communication and LTE system. In: 2015 IEEE 82nd Vehicular Technology Conference (VTC2015-Fall), pp. 1–5 (2015). https://doi.org/10.1109/VTCFall.2015.7390808
14. Wu, X., O'Brien, D.C.: Parallel transmission LiFi. IEEE Transactions on Wireless Communications **19**(10), 6268–6276 (2020). https://doi.org/10.1109/TWC.2020.3001983
15. Popadić, M., Kočan, E.: LiFi networks: Concept, standardization activities and perspectives. In: 2021 25th International Conference on Information Technology (IT), pp. 1–4 (2021). https://doi.org/10.1109/IT51528.2021.9390098

Fuel Larceny and Leakage Indication System Using IoT

Prasanna Palsodkar, Prachi Palsodkar, Aniket Gokhale, Prabhakar Dorge, and Ajay Gurjar

1 Introduction

IoT attracts much attention recently, and it is a technology that brings control of physical devices over the internet. In cutting edge years, IoT technology is advancing the life of people in many ways, and it has more impact on the day-to-day life of the people. Evolution in medical field, agribusiness, digital cities, and digital homes are just a few illustrations where IoT is deeply rooted [1].

In the neoteric years, surging oil demands and cost of fuel are increasing day by day. Security of fuel at global level is still avoided by the people. The count of fuel theft is increased as fuel prices mounting up, appealing more poachers to pilfer the fuel. Poacher can draw off fuel from fuel tanks thereby forging it tremendously difficult to determine that fuel would cautiously be stolen. This enactment is frequently utilized by chauffeurs for their own individual purpose or for merchandise. To overcome this severe issue it is necessary to emerge with such an apparatus that will continuously check the amount of fuel in the reservoir of the automobile and warn the possessor in case of any larceny; there by allowing the possessor to supervise and safeguard fuel from being stolen or prevent leakage from the fuel tank.

Fuel larceny is one of the main bothering issues for numerous automobile owners. The prime motive of this system is to circumvent situation like fuel larceny from the reservoir of the automobile. In suggested system, FGS is used to determine the

P. Palsodkar (✉) · P. Palsodkar · A. Gokhale
Department of Electronics Engineering, Yeshwantrao Chavan College of Engineering, Nagpur, India
e-mail: palsodkar.prasanna@gmail.com

P. Dorge
Department of E&TC Engineering, Yeshwantrao Chavan College of Engineering, Nagpur, India

A. Gurjar
Department of E&TC Engineering, Sipna's College of Engineering & Technology, Amravati, India

S. Mukherjee et al. (eds.), *Intelligent Systems for Social Good*,
Advanced Technologies and Societal Change,
https://doi.org/10.1007/978-981-19-0770-8_7

amount of fuel in the reservoir. If the fuel level goes underneath certain level, then FGS sends a precise signal to the MCU. Then MCU will ON the buzzer and send warning to the possessor of the vehicle instantly. MCU is a prime component or CPU of the system. If the possessor of the automobile is driving then the amount of fuel level will decrease which can activate the MCU through the FGS. To overcome such condition, a signal from ignition key is taken into account. At any moment, the possessor of the vehicle inserts key into the ignition lock and turn it ON then at that instance a signal will be given to the MCU. Through this the MCU identifies that the vehicle has been started and so it will not check fuel level. FGS is turned ON only when the key is removed from the ignition. The system is activated as soon as the key is taken out from the ignition.

Therefore efforts are put-out to design an affordable and constructive system to prevent fuel larceny. The prime edge of this system is that the complete system can be implemented and developed with a modest amount of expenditure. Hence an affordable, durable, and fuel larceny system is designed and developed successfully and realized realistically.

2 Literature Review

Rohitaksha et al. (2014) in their research paper proposed an affordable vehicle discovery and surveillance system using Global Positioning System (GPS) and Global System for Mobile Communications (GSM) to fetch the vehicle location, information and send it to the other static module. They have also implemented the receiver module which accumulates the circulated information by SMS and process it to a congruent format to Google Earth so as to fetch the latitudinal and longitudinal information and status of the vehicle [2].

Aher and Kokate (2012) in their research paper presented the idea of fuel surveillance and vehicle discovery administration system based on MCU. They tried to implement a system which assists the owner of vehicle, seating at remote location, to keep an eye on fuel larceny and discovering the vehicle precisely and interminably [3].

Shubham Kamthe et al. (2020) in their article discussed about measurement of level in fuel tank through android application. The proposed system is able to provide a way for police investigation for larceny or deceitful incidents of fuel larceny or escape. This structure also allows a self-regulating analysis and keeps an eye on fuel level [4].

Komal D/o Shoukat Ali Khuwaja et al. (2018) in their research article proposed the prototype of hardware of fuel control system for automobiles carrying fuel. The proposed system provides Automatic Electronic System (AES) which calculate fuel quantity and sends calculated quantity to the owner's smartphone and computer. The reported system provides a systematic approach for determining larceny or deceitful incidents. It also pinpoints the latitudinal and longitudinal information of each fuel carrying tanker in case of any incident [5].

Obikoya (2014) in his paper presented a system which will keep track of the quantity of fuel in both immobile tanks and moving tanks. He proposed that the system will cut down operating cost and improve surplus for all entity that are using huge vehicle [6].

Hiremath and Patil (2015) presented the idea of fuel recognition system using MCU ARM7. Their system sends text message to user about the precise location of the vehicle. They have used GPS and GSM module which will gives information about the quantity of fuel in the tank [7].

Faizah Farzana et al. (2020) in their paper presented the idea of real time fuel surveillance of vehicle tracking systems. They proposed such a system that incorporates the movement variables of a vehicle in the discovering device to improve the exactness of the quantity of fuel. In addition, they also proposed to ascertain dubious activities and distinguish fuel reloading by administering a time-related analysis of the sustained quantity of fuel [8].

Ganesh Katkar et al. (2017) in their research article presented a system for fuel checking and detection of theft using GSM module. They proposed that when any one tries to steal the fuel then the MCU send the signal to the GSM modem to transmit a text note as a warning to the vehicle owner and to make them aware by alarming a buzzer which is positioned in the system [9].

Pallavi Bendre et al. (2018) discussed about minimizing fuel theft by developing a system which enhances the vehicle alarm security system. Their system will alert the vehicle owner about the fuel theft by sending text message using GSM technology [10].

Chinna Babu and Prakash (2018) in their article proposed a system using IoT for real time tracing and fuel controlling system. The proposed work detects whether the truck is involved in accident or not and send accident location information through email. Their system also provides efficient real time tracking, fuel level, and monitoring truck condition remotely based on IoT technology to protect the truck from theft [11].

Sudharshana et al. (2017) presented their work on automatic alarm system to protect the fuel theft. They have used GSM and GPS technology to transmit SMS to possessor of the vehicle about exact location of the vehicle using the latitudinal and longitudinal position of the vehicle [12].

Johannes Hermanus Potgieter and Patrick Desmond Fourie (2006) proposed a systematic approach for fuel larceny and system which is connected to an automobiles fuel level (FL) sensor for giving an evidence of the FL in the automobile. MCU is used for managing the system. The MCU is used to estimate and stock the median fuel level over consecutive intervals of time. If the fuel level deviates by more than a preset amount, then it predicts fuel larceny. The MCU is concurrently able to ascertain the fuel larceny when fuel tank is being refilled [13].

3 Experimental Work

IoT is enormously used in almost all of the system, and its acceptance is increasing every day. In this article a system is proposed which is designed and developed based on IoT and smartphones or tablet computers for keeping eye on vehicle fuel activities such as fuel larceny and leakage indication. The proposed system administers the quantity of fuel by using fuel level sensor. Level sensor is needed to continuously monitor the reading of the level of fuel in the tank (vehicle tank or any reservoir) and this level sensor is connected to a microcontroller input. IoT is used to implement and monitor the level of the fuel continuously by displaying it on a website using an ESP8266 Node MCU (Wi-Fi Module). There would be a certain level (quantity) of fuel in the reservoir. If the decreasing rate of fuel level is greater than the preset level, an immediate alert notification is sent to the owner through the cloud server.

Fuel Larceny monitoring system consists of Microcontroller, FGS to measure fuel level and Wi-Fi module. LCD display is used to show real time alert, Wi-Fi module is employed to transmit alert to the user about theft and fuel level. When power is ON, all the components of MCU measures fuel level continuously and shows the condition on LCD display whether the ignition switch is ON or OFF. If amount of fuel get reduced during ignition OFF condition then owner get fuel theft alert. Buzzer is used to produce sound signal so that owner of the vehicle will be aware about fuel theft or leakage from the fuel tank.

3.1 Design of Hardware Module

Figure 1 shows the structural flow of hardware module of the system. Here FGS is employed to sense the amount of fuel in the reservoir, and it sends a signal to ADC which is then processed by microcontroller ATMEGA328p. ESP 8266 is Wi-Fi module which is used to send information to the owner about the fuel level in the reservoir and if fuel larceny occurs, and it sends a warning to the user in the form

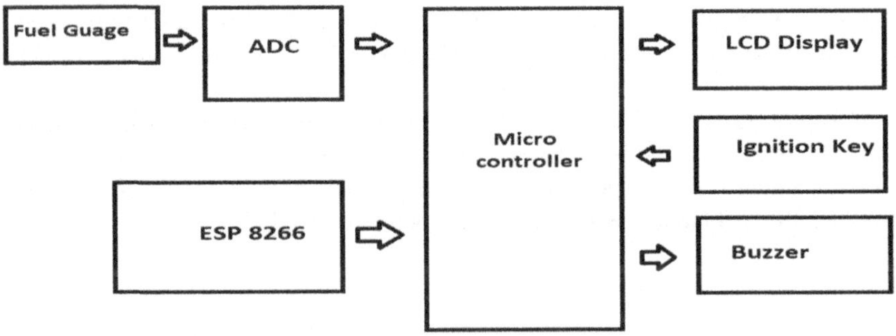

Fig. 1 Structural flow of the hardware

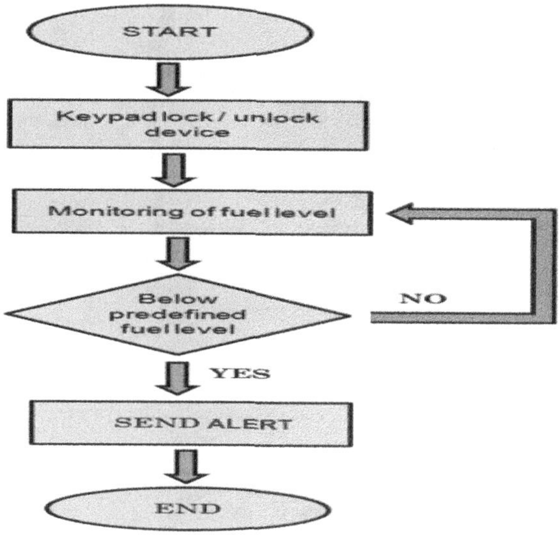

Fig. 2 Flow of the proposed system

of text message. LCD display is used to display the alert sent by the Wi-Fi module. Ignition key module is employed here to represent the state of vehicle whether the vehicle is in moving state or at rest.

3.2 *Flow of the Proposed System*

Figure 2 shows the flow of the algorithm of the proposed system. Initially when the power is ON, then the system will check the condition of the ignition key. If ignition key is in OFF state then it starts monitoring the amount of fuel in the reservoir. For this purpose FL sensor is used and a threshold value is set so that if the FL decreases beneath the specified preset value, then a signal is send to generated which is analog in nature. For converting this analog signal to digital, ADC have been used which is interfaced with microcontroller. The microcontroller will generate an alert signal and using IoT module ESP 8266, it sends signal to the possessor of the vehicle regarding fuel larceny.

3.3 *Proposed System*

The proposed system is extremely easy to operate and can easily be modified according to future scopes. Figure 3 shows the complete system of Fuel Larceny and Leakage Indication System based on IoT. Initially when the key ignition is ON, means vehicle is in ON condition and there is no possibility of fuel theft and therefore

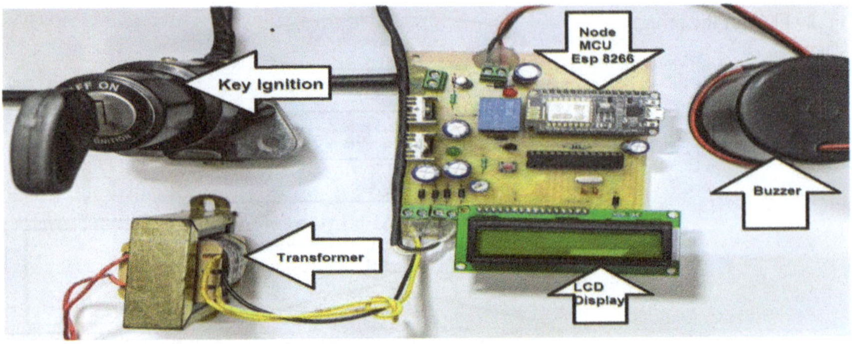

Fig. 3 Fuel Larceny administering circuit (when ignition key ON)

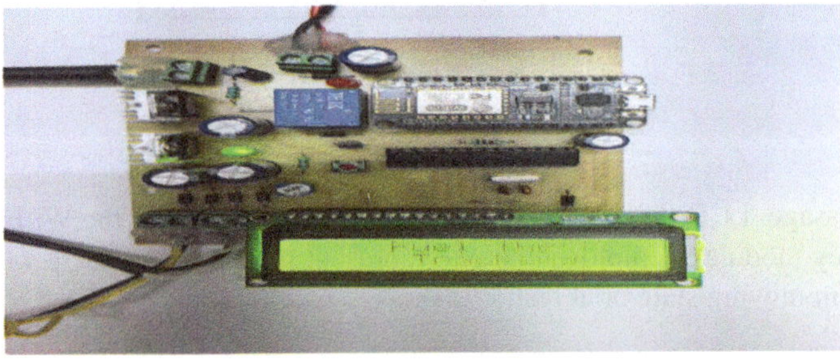

Fig. 4 Fuel Larceny administering circuit (when vehicle in OFF condition)

our system will remain in OFF condition, i.e., LCD will not display any message on it.

Figure 4 shows the system is in ON condition which means key ignition is OFF or when the key was ejected from the vehicle. In such situation, the system will continuously track the amount of fuel in the reservoir and display the message as "Fuel Theft Tracking" on LCD display.

In Fig. 5, the system is in ON condition without any theft or leakage issue. The level of the tank is also at constant level without any variations. Therefore LCD will display "System OK" message on both LCD display and on our web page also.

4 Experimental Result

Figure 6 reveals the condition of the system. When key ignition is in OFF condition and system is in ON condition, if the level in the reservoir decreases due to larceny or leakage from the fuel reservoir, i.e., disparity of the fuel level in the reservoir gets detected by FGS then an alert indication is shown on LCD display and also sent

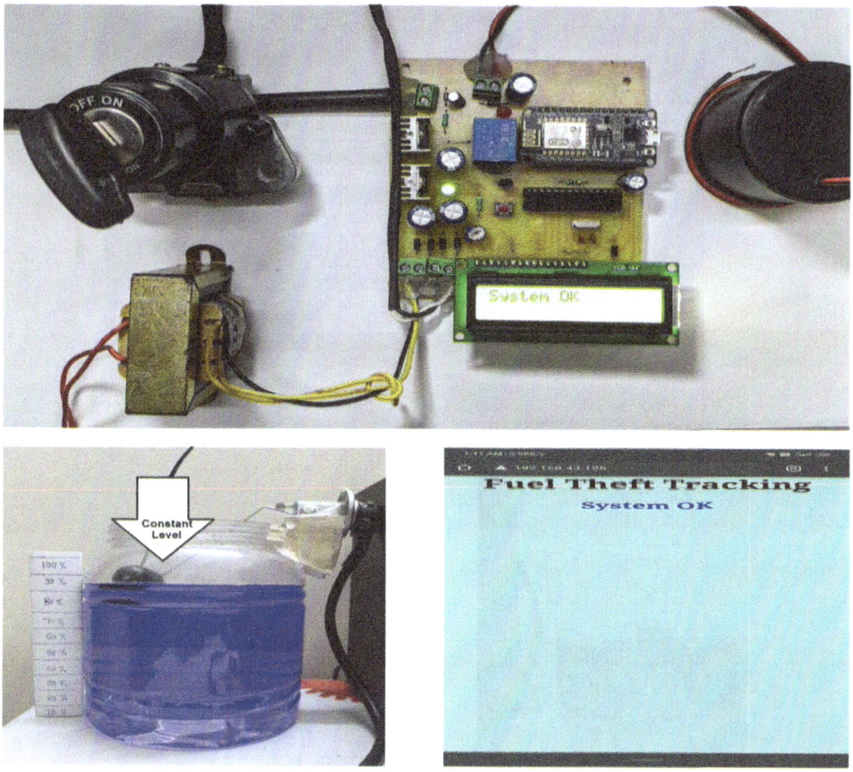

Fig. 5 System condition in ON state

to smartphones and on webpage. For industrial purpose, buzzer is connected in the system which gets turned ON when theft is detected.

5 Conclusion

Detection of fuel larceny is favorably built to explain the detailed logic of larceny detection and leakage indication in the reservoir of the automobile. The need for detection of fuel larceny and leakage indication has become imperative due to escalating prices of fuel. When there is invasion or alteration of fuel in the fuel reservoir, the system becomes operative thereby providing owner the precise indication of vehicle and its fuel content. The prime aim of this system is fuel safety which is provided by the use of Wi-Fi module. An advance and affordable approach for fuel safety is designed and implemented in this proposed work. The proposed system can be established in a small space thereby making it inaccessible. The perceptible feature of the proposed system is that it can consistently transmit the warning until

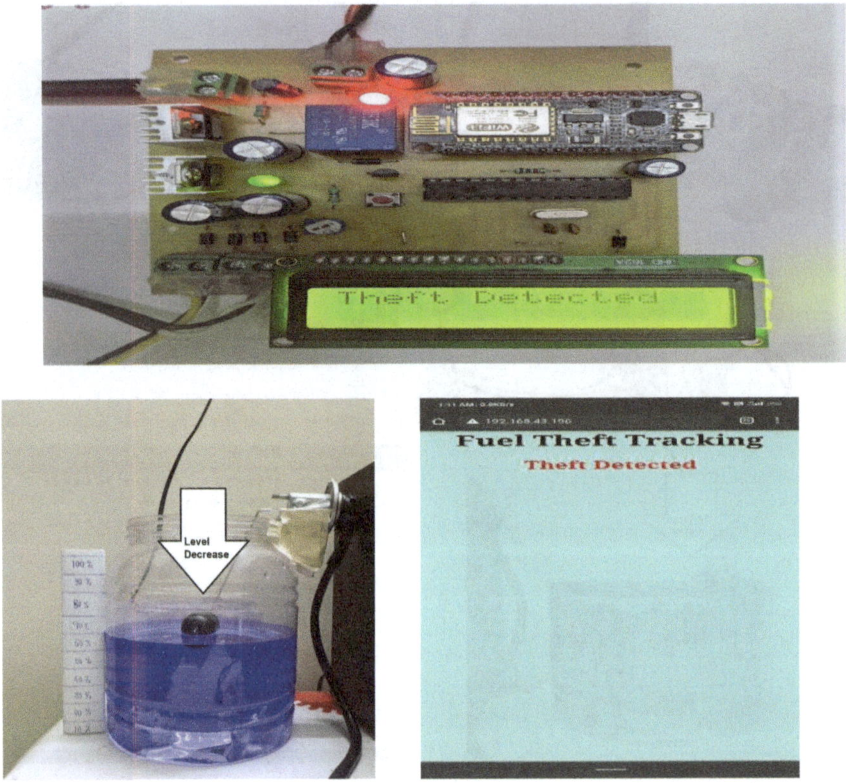

Fig. 6 System showing theft detection

the owner acknowledges it. Many fuel larceny detection models are in use but most of them are either overpriced, reckless, takes lot of space and worthless for transmitting signal over long distance. Presented system is committed to fuel larceny monitoring and detection and is not designed to keep track of any other automobile limitations.

References

1. Sharma, P., Komal, A., Ranjit, N.: Antitheft system for fuel detection using IoT. Int. J. Comput. Sci. Eng. **8**(5), 110–116. E-ISSN: 2347-2693 (2020). https://doi.org/10.26438/ijcse/v8i5.110116
2. Rohitaksha, K., Madhu, C.G., Nalini, B.G., Nirupama, C.V.: Android application for vehicle theft prevention and tracking system. Int. J. Comput. Sci. Inf. Technol. **5**(3), 3754–3758. ISSN: 0975-9646
3. Aher, S.S., Kokate, R.D.: Fuel monitoring and vehicle tracking. Int. J. Eng. Innov. Technol. (IJEIT) **1**(3), 166–169. ISSN: 2277-3754
4. Kamthe, S., Bodkhe, K., Karnavat, M., Shaikh, B., Pawar, S.S.: Automatic fuel tank monitoring

and leakage detection. Int. Res. J. Eng. Technol. (IRJET), **07**(05), 1319–1322. e-ISSN: 2395-0056, p-ISSN: 2395-0072
5. Ali, B.A., Mihalca, V.O., Cătălin T.R.: Automatic fuel tank monitoring, tracking & theft detection system. In: MATEC Web of Conferences, vol. 184, 02011, pp. 1–4 (2018). https://doi.org/10.1051/matecconf/201818402011, Annual Session of Scientific Papers IMT ORADEA
6. Obikoya, G.D.: Design, construction, and implementation of a remote fuel-level monitoring system. J. Wirel. Com. Netw. **2014**, 76 (2014). https://doi.org/10.1186/1687-1499-2014-76
7. Hiremath, N., Kumbhar, M., Pathania, A.S., Patil, V.: Smart fuel theft detection using microcontroller ARM7. Int. J. Eng. Res. Technol. **4**(09), 121–126 (2015). ISSN: 2278-0181
8. Farzana, F., Hossain, M.M., Imtiaze, M.M., Hossain, M.T., Jameel, A.S.M.M., Islam, S.: A real-time motion based fuel monitoring technique for vehicle tracking systems. Emerg. Technol. Comput. Commun. Electron. (ETCCE) **2020**, 1–6 (2020). https://doi.org/10.1109/ETCCE51779.2020.9350860
9. Katkar, G., Lanke, S., Kumar, A., Khawate, S.: Fuel checker and theft indication. Int. J. Adv. Res. Innov. Ideas Educ. **3**(3), 1813–1815. ISSN (O)-2395-4396
10. Bendre, P., Birambole, U., Dinde, I., Shelke, S., Sidhaganesh, S.: Fuel theft protection & indication. Int. J. Innov. Eng. Res. Technol. [IJIERT] **5**(4), 74–76 (2018). ISSN: 2394-3696
11. Chinna Babu, D., Prakash, V.: Real time tracking and fuel monitoring of truck using IoT. Int. J. Pure Appl. Math. **120**(6), 1685–1701 (2018). ISSN: 1314-3395
12. Sudharshana, B., Jacob, S.R., Sukambika, S.: Automatic oil tank truck alarm system for fuel theft prevention. Int. J. Adv. Res. Innov. Ideas Educ. **3**(3), 459–465 (2017). ISSN(O)–2395-4396
13. Potgieter, J.H., Fourie, P.D.: Fuel theft detection system and method U.S. patent application number 20060111851 10/2006-05-25

IoT Applications for Health Care

Shiva Agarwal

1 Introduction

Today, medical care is progressively utilizing data advances for conveying and understanding frameworks pointed toward accelerating wellbeing diagnostics and therapy. Such frameworks offer shrewd types of assistance for wellbeing checking and clinical mechanization in various settings and conditions (clinics, workplaces, home, on-the-go.), [1] hence permitting a significant decrease of doctor visit costs and an overall improvement of patient consideration quality. We all are using smart gadgets much time to monitor our health.

In this unique situation, the wide dissemination of incredible inserted equipment along with the advancement of keen clinical sensors and gadgets for universal medical care has formed the Internet of Medical Things (IoMT) definitely change the manner in which medical services is moved toward around the world, so the quantity of medical care gadgets utilizing Internet of Things and wearable advances, which is required to arrive at before the finish of 2021 [2].

Information caught by wearable (like watches equip with sensors), ingestible and installed sensors, versatility designs, gadget utilization designs permit to follow client propensities and can be viably gathered and handled to uncover basic conditions by utilizing best in class Machine/Deep Learning (ML/DL) and Artificial Intelligence (AI) based methodologies. Customary cloud-based designs for Big Data examination can give great execution and dependability when we are supporting for non-security and inactivity basic of IoT implementation or applications [3].

A developing interest in the designs that understand the collaboration of Fog, Cloud, and Edge registering is of late arising. The fundamental objective is to abuse

S. Agarwal (✉)
National Institute Of Technology, Silchar, Silchar, Assam, India
e-mail: shivaagarwal793@gmail.com

S. Mukherjee et al. (eds.), *Intelligent Systems for Social Good*,
Advanced Technologies and Societal Change,
https://doi.org/10.1007/978-981-19-0770-8_8

the maximum capacity of edge hubs and low-level haze hubs to deal with practical assignments as well as information preparing, investigation, connection, and deduction [3].

Such methodologies address a promising answer toward the execution of solid circulated medical care applications and administrations, since keen planning of computational and asset the board errands across the hubs demonstrates to see the rigid prerequisites of IoMT situation.

In this unique situation, we help to the spread of "Fog/Edge Health" arrangements that utilize reasonably figuring ideal models to disseminate wellbeing sensors information preparing and capacity among various hubs, situated at different degrees of the vicinity to clients, as follows:

- Edge processing happens straight forwardly on the gadgets to whom the sensors are joined or an entryway gadget which is actually near the sensors: instances of edge hubs are wearing gadgets, for example, cell phones, smartwatches, or versatile "specially appointed" installed frameworks like microcontrollers, single-board PCs [4, 5].
- Fog figuring hubs perform at a neighborhood level, they can incorporate greater and all the more remarkable gadgets like PCs, nearby workers and doors that might be genuinely more far off from the sensors and actuators [3].

The two standards (progressively regularly actualized together) influence the vicinity to the client to furnish area mindful wellbeing administrations with decreased inertness and high accessibility. A few techniques depending on progressive figuring methodologies have been given to dispense and convey the induction errands of ML and AI strategies in the middle of the cloud, the haze, and the edge levels, attempting to raise the (restricted) arithmetic limits of edge gadgets to their extreme [4].

A change against the versatile mobile cloud computing model (MCC), described by big information communication expenses and restricted inclusion toward a mobile edge computing model (MEC) [2] with down-idleness and solid edge ML models, is then dynamically occurring in the shrewd medical services space.

We perform an audit about IoMT arrangements, zeroing in on wellbeing observing. We mean to show the advancement of IoT-based medical care frameworks, beginning from observing arrangements, that include edge hubs just for practical undertakings, up to late recommendations that influence agreeable edge/haze figuring for brilliant wellbeing. We along these lines examine patterns in Edge ML procedures and models [6].

2 Literature Survey

Liu [7]: Year of Publication: 2017 This exploration investigates uses of IoT for reconnaissance and checking that is utilized in consistently life. These incorporate arrangements like security observation, medical services, autonomous living, and so forth to address the issues seen because of information caught from different perspectives and heterogenous sensors used in IoT, the creator has given a novel strategy for class-compelled move direct examination. This technique helps in extraction of invariant highlights from the caught information. The examination work on cross view activity acknowledgment of IOT checking frameworks and proposes a model that takes care of the issue of "human activity" recognition because of pictures caught at various points utilizing the capacity of extraction of highlight invariant measurements. The exploratory outcomes have exhibited that the given CTLDA can accomplish preferred outcomes over the best-in-class strategies [7].

Pinto et al. [8]: Year of Publication: 2017 Because of an increment in the number of inhabitants in the matured across the world there has been a developing necessity to give arrangements that give living help to the old populace. In this viewpoint in may be said that the IoT can give another perspective to present day medical services by giving a more customized, preventive and shared type of care. This examination work presents a living help based IoT answer for the older that can screen and enroll patient's crucial data just as give systems to trigger cautions in crisis circumstances. The examination work proposes an answer including a wrist band or the smart watch that can associate with the cloud worker to screen and help old individuals. It professes to be low force/cost arrangement with Wireless correspondence abilities.

3 Structure for IoMT System

The utilization of Edge/Fog registering models in the medical services space typically manages the plan of distant observing arrangements that influence wearing and field sensor networks for executing defensive, preventive, and responsive frameworks [9, 10]. In this unique situation, most commitments include fog hubs going about as nearby workers that gather and cycle wellbeing information to react to the help prerequisites rapidly. Mainstream researchers have been examining for over ten years answers for checking patient's wellbeing status with the point of distantly giving reports to clinicians.

We basically propose PC-based frameworks to distantly screen patients, for instance, utilizing information from accelerometers and ECG to instruct clinicians about periods with respect to raised pulse and channel anticipated basic circumstances. Ensuing executions use miniature regulators and one-board PCs to assemble emotional information from specific sensors (E-wellbeing sensor stage, ECG) to be additionally handled and broke down in a PC-based climate.

As of late, the development of IoT innovations has made ready to the advancement of a few more astute arrangements abusing both organizational structures and programming stage areas. Such arrangements expect to address medical services at various levels: pediatric and old consideration, persistent sickness oversight, plague illness observing, clinical digital actual frameworks, personal wellbeing, and wellness the board.

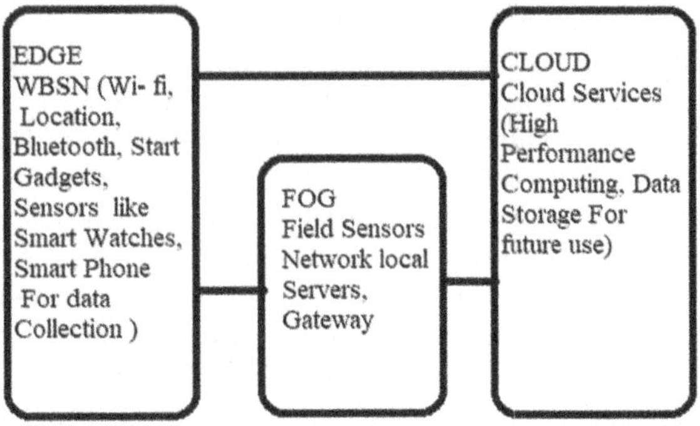

Fig. 1 Three tier architecture healthcare model which based on IoT and edge computing

In this work, we primarily center on frameworks tending to wellbeing checking issues that can be likewise part in two classes: static distant observing, where the patient should be in a structure (hospital, home), and dynamic checking, which expects the patient to be checked in a hurry.

An overall arrangement utilizes a staggered design, appeared in an extremely worked on the route [11] in Fig. 1 that should include:

- An Edge level, where versatile gadgets (cell phones, smartwatches, minimized inserted frameworks) carry out pre-preparing and some down-level explanation on information gathered from Wireless Body Sensor Networks (WBSN).
- A Fog level, where Personal computers, workers/entryways accumulate information from field sensor organizations to perform neighborhood handling or potentially to stockpile.
- A Cloud level, where cloud administrations are known for elite registering assignments and distant information stockpiling.

The three levels do not need to be wholly actualized. For instance, in static checking issues, information from sensors must be straight forwardly gathered by fog gadgets and be expounded with discretionary help from cloud administrations. Similarly, some powerful observing situations where a fog level could not actualize, edge gadgets must be straight forwardly cooperating with cloud administrations [8].

This segment focuses on the best way to shrewdly utilize and focus on organization assets in an IoT structure focus on a protected and believe commendable transmission path for a medical care-based application. This can be accomplished by early processing the info information got from the detector systems at the end gadgets. For this, the end gadget may take the help of the backend cloud workers that approach weighty processing assets. The back end can play out the weighty calculating and exhort the endpoint gadget in regards to the particular preprocessing to be done to have the option to focus on approaching information from sensor structures [2]. The backend can utilize ML and information mining ideas to separate marks from coming sensor information and appropriately give clinical translation dependent on the caught information. Using this frontend gadget can give an evaluation of the patient medical issue.

The execution can be reached out to give a technique so just conspicuous vacillations can be given to the back-end where a doctor can examine the information and finish up. This helps in distant conclusion and give better/best country drug where the specialist is away. It has been seen that because of organization latencies, cloud figuring does not find a way into territories that require ongoing low dormancy reactions. This is fundamental because of the great idleness of choices. This has prompted another disseminated processing engineering proposition as "edge registering," where small piece of the calculations should be possible at the IoT gadget or "edge" [5] gadgets instead of having everything figured in the cloud. The essential focal point of this examination has wed the ideas of cloud registering and IoT along these lines centering around upgrades in edge figuring strategies chiefly for the medical care area.

3.1 An IoT Based Endpoint

- Has various detectors associated with it.
- An IoT finish point is associated with a backend worker over a Wireless organization
- The IoT based finish point can speak with the IoT Back-end through IoT conventions.

3.2 The Backend Worker

- Can perform calculation escalated errands as, it approaches the top-of-the-line figuring assets.
- For calculation assignments, the backend can utilize various methods like massive information investigation.

3.3 AI, Neural Organizations

- Can be cloud-based. Be that as it may, our emphasis would be tense, figuring
- Can send notices to the IoT Endpoint.

Methods that we can utilized for this execution.

- IOT convention

A customer would be running on the IoT based End-point and an IoT conversation worker would be executing in the cloud (Edge Server) as shown in Fig. 2. This will give the vehicle convention important to correspondence between the IoT Endpoint and the information worker.

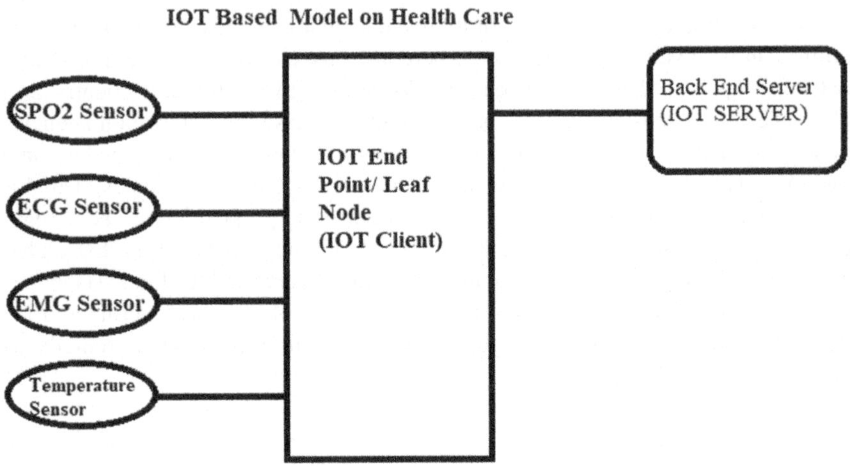

Fig. 2 IOT model for health care

As examined before, because of high organization latencies cloud registering is not thought of and an ideal answer for Real-Time Data Analysis [10]. For this reason, we have Edge-Computing design which has been presented. Our fundamental thought can be stretched out to edge figuring design where the end-point gadget can achieve calculation dependent on direction from a leaf point gadget which thus can get information from Cloud workers.

4 Conclusion

IoT answers for medical services are advancing from straightforward structures to gather, send, and imagine information gained by field and wearing detector networks toward complicated brilliant frameworks ready to give examination, perceive exercises, and deciding [12]. In this progress, AI and AI strategies assume a significant part yet their usage requires a computational limit that is frequently accessible exclusively by methods for cloud administrations. In reality, with the dramatic development of the information created by the detector, a cloud-driven vision of the ML preparation current a few shortcomings for various goal. The nature of the assistance is firmly impacted by the nature of the web association, which additionally represents an accessibility issue. Specifically, where an ongoing exhibition should be guaranteed, as in early-identification, hazard anticipation, or action acknowledgment, the reaction time should be low to permit medical services suppliers to proactively respond to conceivable corruption of ailments. Information stockpiling and security are additionally basic issues when managing wellbeing related administrations because of the enormous measure of individual information to be overseen. All things considered, the decision of a completely nearby administration is as yet unreasonable because of impediments in handling force and capacity, particularly on account of dynamic checking. A few propositions that attempt to plan a legitimate figuring model to the IoT levels are being acquainted with encounter the necessities of IoMT frameworks. They target conveying the DL responsibility in middle of the cloud and fog hubs, regularly consigning the model preparing to the cloud and moving the dynamic to the edge [6]. The profit in these methodologies is very much persuaded by the dispersion of minimized GPU implanted equipment which permits the sending of ground-breaking and compelling fog hubs. In situations where the DL preparing stage is still computationally cannot be managed by the edge hub, DNN apportioning and conveyed preparing addresses a promising pattern as of now.

References

1. Abdellatif, A.A., Mohamed, A., Chiasserini, C.F., Tlili, M., Erbad, A.: Edge computing for smart health: context-aware approaches, opportunities, and challenges. IEEE Netw. **33**(3), 196–203 (2019)
2. Akmandor, A.O., Jha, N.K.: Smart health care: an edge-side computing perspective. IEEE Consum. Electron. Mag. **7**(1), 29–37 (2017)
3. Azimi, I., Anzanpour, A., Rahmani, A.M., Pahikkala, T., Levorato, M., Liljeberg, P., Dutt, N.: HiCH: hierarchical fog-assisted computing architecture for healthcare IoT. ACM Trans. Embed. Comput. Syst. (TECS) **16**(5s), 174 (2017)
4. Abu-Elkheir, M., Hayajneh, M., Ali, N.A.: Data management for the internet of things: design primitives and solution. Sensors (Basel) **13**(11), 15582–15612 (2013). https://doi.org/10.3390/s131115582. Published online 2013 Nov 14. PMCID: PMC3871070
5. Hassan, N., Gillani, S., Ahmed, E., Yaqoob, I., Imran, M.: The role of edge computing in Internet of Things. IEEE Commun. Mag. **56**(11), 110–115 (2018). https://doi.org/10.1109/MCOM.2018.1700906
6. Ai, Y., Peng, M., Zhang, K.: Edge computing technologies for Internet of Things: a primer. Digit. Commun. Netw. **4**(2), 77–86 (2018). ISSN 2352-8648
7. Liu, S.: Class-constrained transfer LDA for cross-view action recognition in Internet of Things. J. Latex Class Files **6**(1) (2017)
8. Cabral, J., Gomes, T.: We-Care: An IoT-Based Health Care System for Elderly People by S. Pinto, pp. 1378–1383. IEEE (2017)
9. Greco, L., Percannella, G., Ritrovato, P., Tortorella, F., Vento, M.: Trends in IoT based solutions for health care: moving AI to the edge. Pattern Recognit. Lett. **135**, 346–353. https://doi.org/10.1016/j.patrec.2020.05.016. Epub 2020 May 13. PMID: 32406416; PMCID: PMC7217772
10. Alwan, O.S., Rao, K.P.: Dedicated real-time monitoring system for health care using ZigBee. Healthc. Technol. Lett. **4**(4), 142–144 (2017)
11. Al-Fuqaha, A., Guizani, M., Mohammadi, M., Aledhari, M., Ayyash, M.: Internet of things: a survey on enabling technologies, protocols, and applications. IEEE Commun. Surv. Tutor. **17**(4), 2347–2376 (Fourthquarte) (2015)
12. Uddin, M.Z.: A wearable sensor-based activity prediction system to facilitate edge computing in smart healthcare system. J. Parallel Distrib. Comput. **123**, 46–53 (2019)
13. Ahamed, F., Farid, F.: Applying Internet of Things and machine-learning for personalized healthcare: issues and challenges. In: 2018 International Conference on Machine Learning and Data Engineering (iCMLDE), Sydney, NSW, Australia, pp. 19–21. https://doi.org/10.1109/iCMLDE.2018.00014

Smart Healthcare Remote Monitoring System Using Internet of Things

Shyamapada Mukherjee, Rahul Kumar, and Sumanta Banerjee

1 Introduction

Health has the highest importance in a person's life. To lead a good life we need to have a good health condition. Unfortunately, due to unavailability of doctors, infrastructure and other reasons, some of us can not get the right facility at right time. Here, the IoT based healthcare instruments prove to be revolutionary. Which can provide efficient and imminent information about a person's health. The information may include heart rate, body temperature, and body position. Which will be sent to the patient's doctor via email. IoT can be defined as a network of sensors and actuator devices, that can sense and transmit data, and manipulate the physical environment without direct involvement of a human. This definition may be insufficient to give a clear picture of the technology as it is still in its nascent state. All its components such as the devices, the adhoc network technology are evolving towards providing best services. Healthcare has also gained several smart and useful systems that has improved the services in terms of efficiency and accuracy [1]. Sitting posture identification, automatic tracking of people and bio-medical devices in hospitals and clinics, correct drug-patient associations, real-time monitoring of patient's physiological conditions for the best treatment are only a few of the possible examples. Researchers across the world are engaged in developing more sophisticated and accurate devices (such as nano-scale devices that can be deployed inside human body for better diagnostics) for IoT enabled healthcare systems. So, this technology has all the potent to offer timely and lifesaving information service to us and the possibilities are ever increasing. Sitting analysis becomes an essential part of the health care system. People driving four-wheeler vehicles, students especially while studying in the classroom, employees, and staff working in the companies spend more time sitting in their daily life activities. It is often found that, students in a class keep sitting in inappropriate pos-

S. Mukherjee (✉) · R. Kumar · S. Banerjee
National Institute of Technology Silchar, Silchar, Assam 788010, India
e-mail: shyama@cse.nits.ac.in

S. Mukherjee et al. (eds.), *Intelligent Systems for Social Good*,
Advanced Technologies and Societal Change,
https://doi.org/10.1007/978-981-19-0770-8_9

tures. This habit of spending long hours in this way results in many bad side effects like neck pain, low back pain, etc. It may result in sciatica and cervical spondylosis, if the same habit continues. Thus, there is a need for some efficient solution in areas of sitting posture analysis. Research studies showed great concern to solve the problem of sitting posture classification in the health-care domain, especially for elderly, diabetes and cognitive disorders suffering people. The chair, which is available today, is a passive object that does not sense to user's response. The traditional method is to let the patient sit on a hospital chair and then his/her sitting posture is analyzed by observing the sitting behavior to identify the posture. This diagnosis generally takes an hour or a half, although it is not enough for an accurate diagnosis. Doctors are always interested to get those solutions that deliver reliable diagnosis for overall information about the patient. In our system, we designed the sensor sheet for prediction of the sitting posture. Although there are two basic methods of sitting posture recognition which are discussed in different related articles. Those are (1) wearable sensors that a person can put on her body or clothe, (2) another type follows posture recognition from captured images. The latter approach has privacy issues, this is not much practical to use. In recent years, health care sensors along with Arduino, Raspberry Pi and similar tools play a vital role. The Raspberry Pi, an affordable and handy single-board computer that has a set of GPIO pins to connect to different electronic devices. These pins allow it to control these devices, explore IoT and, do physical computing. Wearable sensors are in contact with the human body and facilitate a person to monitor his or her health parameters. The proposed pressure sensor sheet can be put on the driver's seat while driving vehicles. In our project, we are measuring the patient's parameters (temperature, heart rate, and body position) with different available sensors in the market. Among these, pressure sensors have importance in sitting posture identification. In our proposed system, we designed a pressure measurement sheet with only five sensors to determine a person's sitting posture. After successful completion of the experiment, if the specified posture condition satisfied, then we collect the patient's heart rate along with other medical readings like body position, and temperature from the sensors. The function of the proposed system is to store and present the processed information on the Apache-MySQL server. The physician of a particular patient can go through these information and prepare a report. This proposed system provides real-time visual representation while acquiring data. The other key advantages of this prototype are the low cost of the modules and their low power profile. This article is written in the following sections, the Sect. 2 gives a Literature Survey. In Sect. 3 the proposed system has been discussed with the help of images and flowcharts (including design methodology and system architecture). Section 4 introduces experimental results and finally, Sect. 5 gives the conclusion.

2 Literature Review

Budida et al. [2] proposed an IoT-based health care system that collects all the data like temperature, pulse rate and blood pressure using sensors, related to a patient. Then conveys the data to the Wamp based database server, being connected to the patients. The doctor and the patient both can view this medical information via webpage and android applicaion. Which may be used for patient's further analysis and tracking.

Sivagami et al. [3] proposed The Smart Hospital System (SHS) providing monitoring and tracking of patients, personnel, and bio-medical devices automatically within healthcare institutes depending on a CoAP, 6LoWPAN, and REST paradigms. The system offers a mobile-based application that gives faster and better services in medical emergency cases. The data collected from sensors are transferred to a processing and control center, where they are made available by both local and remote users with the help of a customized web service. Validation of the SHS has been performed usingt two different use cases. The first deals with monitoring of the and, the second deals with the management of emergency situations caused by fall of a patient detected by the three-axis acceleration measurements.

Gupta et al. [4] has presented an informative review of the applications and various existing technologies in the healthcare system. A health monitoring system was designed which is smart enough to monitor the patient automatically using IoT techniques. It collects the health data including patient's heart rate, temperature, and ECG and sends the information having all the current medical updates, and emergency alerts to the doctor.

Natrah et al. [5] presented a sensor based measurement system for sitting posture detection. It has experimented the concept with using four pressure sensors and, the results were obtained for three subjects with weights 42kg, 50kg, and 78kg. The efficiency of 8 sitting postures is shown as a line graph. The sensors placing was done in the hip and back seats, which took continuous readings from the sensors. It achieved an overall accuracy of 90%.

Xu et al. [6] presented a concept that requires textile-based sensors, called the eCushion, used to monitor sitting postures. Naive Bayes network method is used to train the data collected from the sensors and selected sensors that are significantly featured for classification. Although this method is dependent on the given training data rather than uncertain parameters like weights, sizes, and sitting orientations. Other techniques were also applied to improve the results of sitting postures recognition, including sensor calibration, dynamic time wrapping-based classification, and data representation. After performing several experiments, the last experiment results showed the classification rate in excess of 85.9%. Basically, the outputs of the sensor sheets doesn't depend only on the sitting posture, in fact, the sensor value depends also on the other uncertain factors like offset, crosstalk, scaling and rotation effects.

Sharma et al. [7] proposed a smart health monitoring system to monitor the health situation of a patient and screen it to the doctors and other paramedical staff through the IoT. The system monitors patient's current status irrespective of the geographic location of the doctor. This paper concentrates on measuring the important parameters of health such as ECG, heartbeats and blood pressure altogether on a single kit, which are transferred to the server of the database and displayed in the website that can be accessed only by authorized personnel. After analyzing the data doctors can then prescribe the medication based on the health report shown by the system, hence this prototype minimizes the burden to visit the doctor every time. the system even provides the facilities to view the patient's previous history, add more patients, and so on.

Kamiya et al. [8] developed a sensor sheet to experiment with sitting posture and also assessed the sheet's performance in two different cases. First, a person-unknown case with a classification accuracy of 93.9% and, second, person-known case with classification accuracy improved to 98.9%. The 5% difference shows the difference of person's weight, position, and other uncertain factors. In the experiments, actually, nine postures were classified that includes leaning forward/ backward/right/left, etc. along with the time series behavior of pressure force.

Archip et al. [9] proposed a prototype system for remote monitoring of low-power sensor arrays that are dedicated for EKG, SpO2, temperature and, body position. The main aim of developing this prototype is to fill the gap with visiting the doctors' clinic every time. The system is composed of sensors with low-power consumption and, that communicate wirelessly with a gateway using RESTful based web interface implemented using the Raspberry PI B+ board.

Huang et al. [10] A smart chair was built by the authors to detect the sitting posture. Eight sitting postures were shown for overall classification using ANN classifier. In this paper, MATLAB was used for the implementation of ANN classifier to recognize the sitting behavior of a person. It yielded high classification accuracy, but focused on only static sitting postures. However, the objective of this project was to present a monitoring system that works for both static and dynamic postures.

Vadivel et al. [11] proposed human activity monitoring system using piezo-electric based accelerometer sensors. the proposed framework was validated by different algorithms and was modeled using WEKA. This paper showed that the estimation of human step length varies linearly in depending on the frequency of walking and the accelerometer variance, and obtained the 97.9% accuracy in step length calculation.

Satija et al. [12] presented a good quality ECG telemetry system which IoT-enabled and consists of three modules(sensing module, assessment module; and analysis and transmission module) for the monitoring applications of heart patients. This quality-aware IoT paradigm has potentials in the application of assessing clinical acceptability of ECG signals. Which may have good application in the improvement of accuracy and reliability of the diagnosis systems that depend on the morphological features and RR interval features for machine learning approaches. The overall evaluation results were taken from MITABIHA for the ECG segments. The results were acquired for twenty subjects, showing the accuracy of more than 98% for noise-free and, 99% for noisy ECG signals.

3 Design Methodology

The proposed system is divided into two subparts: (1) Sitting posture analysis (2) Health care monitoring; that are combined to provide efficient results about patient's health using IoT technology. The main aim of the proposed system is continuous observation of the patient's health status over the local server connected within a specified range. The overall system is subdivided into first, hardware components and, second, software components. Thus, the framework for smart health care remote monitoring system using IoT leverages three-tier architecture, namely: (1) Experiments using Pressure sensors (2) Data acquisition about health parameters from sensors; and (3) GUI based web page design. As shown in Fig. 1, illustrating the overall model, that includes an intelligent health care remote monitoring system that can keep track of the patient's health automatically and efficiently using IoT. This would make the doctor's work easy in looking after the patient from anywhere and any place, as well as the patient can access the health status in the same way.

The diagram is divided into two parts as mentioned above.

3.1 Hardware Components

The proposed design has a sensor sheet that measures pressure force after a person seats on it. The construction and sensor positioning are shown in Fig. 2. There are 5 pressure sensors fitted on the sheet out of which 3 are dedicated for the back side and 2 for the front side. As per the specification given in Table 1 the sensors are very thin and can be easily bend when a user sits on the seat with this sensor fitted sheet. Due to its light weight it can be carried anywhere and used on almost any chair, the user is comfortable with.

Here, our focus is only the static postures, i.e, when the human body postures reaches an equilibrium state. For the experiments, subjects were asked to perform 8 standardized sitting postures : (P1) No Sitting, (P2) Straight, (P3) Forward lean, (P4) Backward lean, (P5) Left lean, (P6) Right lean, (P7) Right leg over left, (P8) Left leg over right, which is shown in Fig. 3 and its description in Fig. 4. 50 trails for each sitting posture (50 $\times$ 8 postures) 450 frames were collected for each subject. Those who have volunteered (subjects) in this experiment were 9 college students weighing 52–64 kg and having age between 21 and 24 years. Each person was first requested to sit down in each posture position and return to the normal position after each posture. There was a break of 2–3 s between two different postures. Thus, in total 3600 training data were collected from pressure sensors.

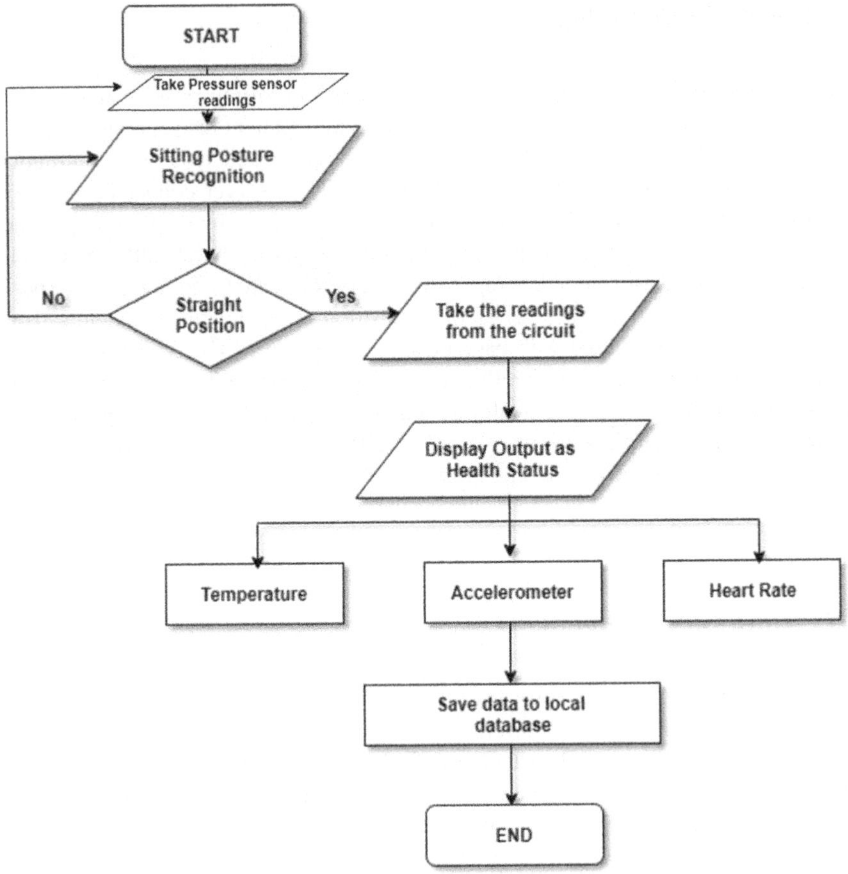

Fig. 1 Block diagram of the whole system

Table 1 Pressure sensor specifications

Thickness	0.208 mm
Length	203 mm
Width	14 mm
Strain sensitivity	5V/$\mu\epsilon$
Threshold	0.00001 $\mu\epsilon$
Span to threshold ratio	100,000,000
Sensing area	9.53 mm

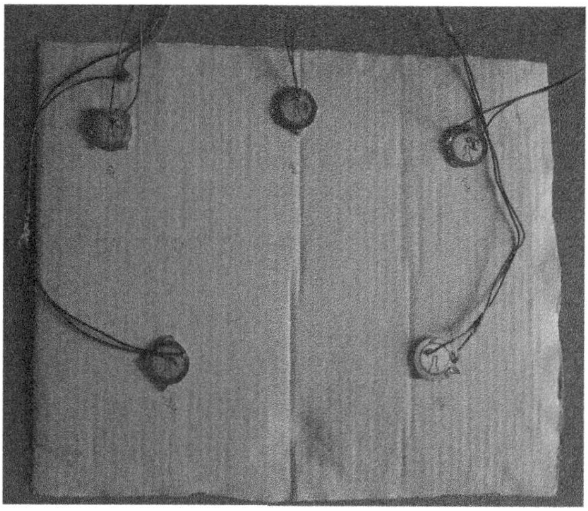

Fig. 2 Sensor sheet construction

3.1.1 Data Collection for Sitting Posture Analysis

The focus of this proposed system is on static postures, that is, the body posture when it ends up at a balanced state. In these experiments, the volunteers were asked to perform 8 standardized sitting postures : (P1) No Sitting, (P2) Straight, (P3) Forward lean, (P4) Backward lean, (P5) Left lean, (P6) Right lean, (P7) Right leg over left, (P8) Left leg over right, which is shown in Fig. 3 and its description in Fig. 4. 50 trails for each sitting posture (50 × 8 postures) 450 frames per volunteer were taken. There were 9 volunteers, who are college students aged between 21 and 24 years, having weights between 52 and 64 kg. Each person was first sat on the chair in each posture position and return to the normal position after a posture. Between each posture there was a 2–3 s break. Thus, in total 3600 training data were collected from pressure sensors.

Figure 5 represents the correlation matrix that shows correlation coefficients between sets of features.

The Fig. 6 shows the probability distribution and relationship for each pair of features in the Data set.

3.1.2 Data Collection for HealthCare Monitoring

This is an integrated system that is a fusion of hardware (sensors, Raspba) and software components. The components related to hardware include Microcontroller ATmega 2560, Raspberry Pi 3 model, Temperature sensor, Heartbeat sensor, and 3-axis Accelerometer.

We have also built a prototype system which includes the following sensor nodes:

About Data

Our DataSet has 5 features, names are given below:

```
Feature 1:        Sensor1
Feature 2:        Sensor2
Feature 3:        Sensor3
Feature 4:        Sensor4
Feature 5:        Sensor5
```

And the class label is "S_Pos", there are total 8 classes given below:

1. No Sitting,
2. Straight,
3. Forward Lean,
4. Backward Lean,
5. Left Lean,
6. Right Lean,
7. R_F_O_L ---------->Right_Foot_Over_Left
8. L_F_O_R ---------->Left_Foot_Over_Right

Total No of Data Points are : 3599

```
 CLASS Label           No_of_Related_Data_Points
Forward Lean:           500
Straight:               500
Right Lean:             500
R_F_O_L:                500
Backward Lean:          500
Left Lean:              500
L_F_O_R:                500
No Sitting:              99
```

Fig. 3 Data collected from pressure sensors and assignment of labels

	Sensor1	Sensor2	Sensor3	Sensor4	Sensor5
count	3599.000000	3599.000000	3599.000000	3599.000000	3599.000000
mean	117.055571	93.449291	135.209225	318.773270	48.625174
std	264.514289	209.399113	278.104018	343.169786	95.089001
min	0.000000	0.000000	0.000000	0.000000	0.000000
25%	1.000000	0.000000	2.000000	41.000000	1.000000
50%	9.000000	4.000000	17.000000	90.000000	5.000000
75%	84.000000	28.000000	92.000000	548.000000	42.000000
max	1023.000000	989.000000	1000.000000	973.000000	419.000000

Fig. 4 Data description

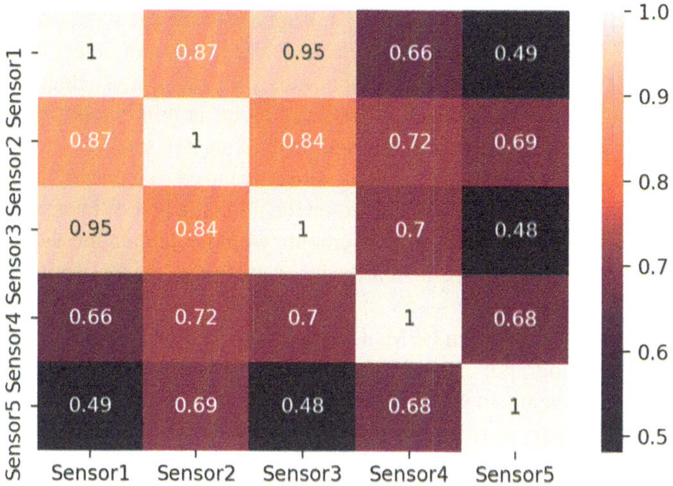

Fig. 5 Correlation matrix

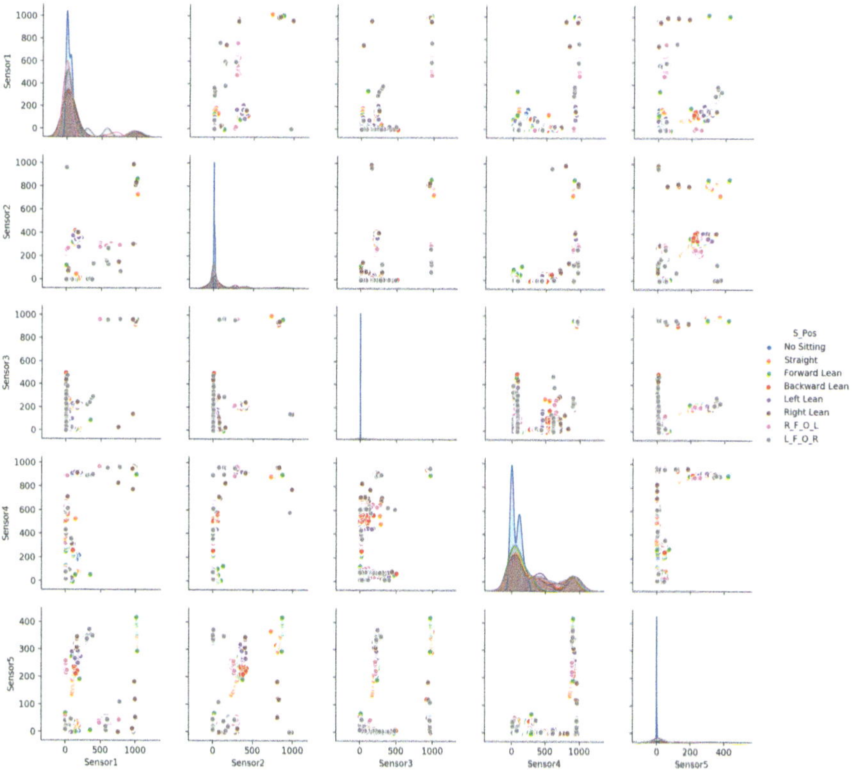

Fig. 6 Pair plot

- Temperature and accelerometer nodes: For temperature, we have used an LM-35 temperature sensor to measure skin temperature, which can detect cold, fever or sore throat. This sensor measures temperature more accurately than thermostats, and also it does not require output voltage to be amplified. ADXL335 3-axis accelerometer is used to detect abnormal movements like seizures, excessive shivers or falls. In this sensor, X, Y, Z -axes values are used to measure the movement of patient's body, which can be installed on the patient's bed. Whenever the patient moves, the sensors observing the movements would calculate the values.
- Heartbeat node: The sensor is a Heartbeat sensor (pulse sensor) which is attached to the patient. It takes +3.5V – +5V at VCC, signal pin to A0 pin Arduino ATmega board for getting heartbeat in BPM. The heartbeat sensor is based on the principle of photoplethysmography (blood flow through the finger at each pulse). Here, we have done experiments on two types of heartbeat sensors: reflective and refractive. When the heartbeat is working, the beat LED flashes in unison with each heartbeat, whenever the finger is placed on it. These readings can be used to detect normal heart condition, chest pain or other heart-related issues.

3.2 Software Components

The Raspberry Pi configuration uses I2C (accelerometer) and 1-Wire (temperature sensor). After the nodes upload the start-up code on Arduino IDE, it tries to join the serial communication with Raspberry Pi, which sends all the data to Wamp based database server to log the patient timely record or timely sensed data, which will help the doctor in generating health report.

Nodes did not perform any specific processing (i.e. data is not interpreted) at the Arduino IDE Compiling level. The web server module provides an interface for real-time data access. Moreover, these data stored in the database for each sensor is shown in tabular form in Fig. 11. The server has an option for uploading the patient's details also. The doctor and the patient can access the data server as and when they want to check the current medical status.

4 Experimental Results

4.1 Results for Sitting Posture Analysis

In the first part of the paper, the training and testing data collected were of different subjects. Then, the training data was split into X_Train, X_Test, Y_Train, Y_Test with test size = 0.33 using Jupyter Notebook App. Jupyter Notebook is a client-server application that allows editing and running python codes via a web browser. It can be executed on a local desktop which runs in a stand-alone system, or can

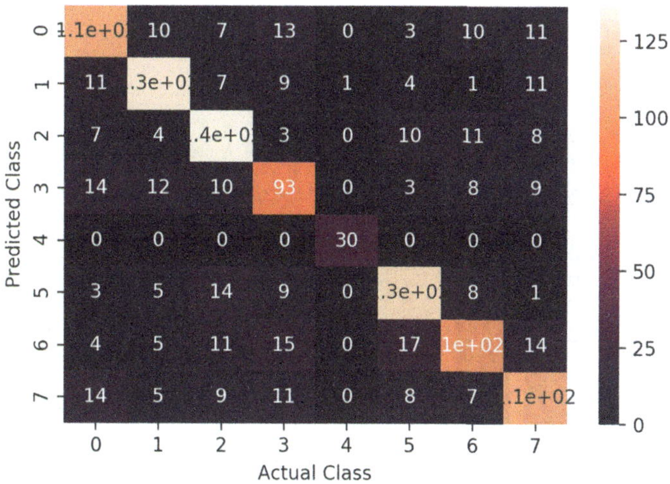

Fig. 7 Confusion matrix

Fig. 8 Classification report

	precision	recall	f1-score
No Sitting	0.56	0.76	0.64
Straight	0.72	0.80	0.76
Forward Lean	0.69	0.74	0.72
Backward Lean	0.68	0.61	0.64
Left Lean	0.94	1.00	0.97
Right Lean	0.79	0.77	0.78
R_F_O_L	0.78	0.61	0.68
L_F_O_R	0.82	0.65	0.73

be installed on a remote server and accessed through the internet. We repeated such process for all different classifiers and obtained the average accuracy for each. The average accuracy using K-Neighbors Classifier obtained is 70.79% using 10-fold cross validation.

The experimental results show that the overall accuracy of the classification of overall sitting positions can be improved. A larger number of sensors will usually result in higher accuracy, but also yield a higher cost. In the future, we will try to improve the rate of recognition with more sensor addition and the experiences we learned in ML field.

The Fig. 7 shows the confusion matrix that describes the performance of the classification model.

With the properly trained KNN classifier, although overall classification accuracy of eight sitting postures obtained was 70.795%, it was found that the classification rate of sitting-upright (P2) achieved is 78% which is the optimal result (Figs. 8 and 9). It can be observed that postures P3, P4, P7, and P8 showed classification rates up to nearly 78%. It may be due to the pressure patterns of these postures are quite similar,

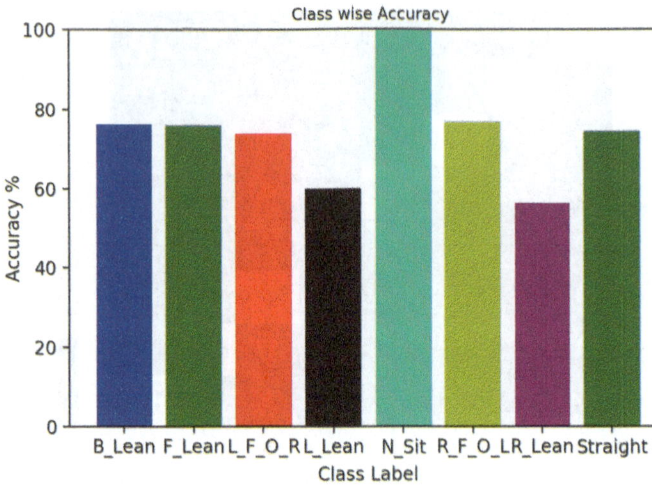

Fig. 9 Class wise sitting positions accuracy

because the hip keeps relatively fixed in these postures. The classification rates of other postures P5 and P6 were close to 60%. These accuracies of each position can be improved to nearly 99.989% using more sensors in the hip seat as well as the backseat.

B. Results for Health care Monitoring System

The health monitoring system introduced in this paper is tried and tested on a normal person, whose personal details are entered into the web portal and stored on the MySQL database server. The way of working of the system is explained by the figures given below, which include phpMyAdmin database connected to the dynamic web page. This page uploads the received data by the sensors and shows a health report.

The home page of the web portal consists of various functional tabs including patient login, About, details of Registration yourself, Logging into your account and changing your password (Fig. 10).

In the patient login page, the user enters the patient id and password, which can be later changed by the user (Fig. 11).

After successful login, the patient can view his last updated health status on his profile page (Fig. 12).

The doctor can log in to his page with the unique user ID and password as shown below Figs. 13 and 14.

Patient registration is being done by the doctor after gathering the data from the sensors. The doctor can manage the patients including patient registration, add new patient, delete patient, etc. and the system systematically maintains the records (Fig. 15).

The patient registration page allows entering the personal details such as Name, Age, Gender, E-mail and so on (Fig. 16).

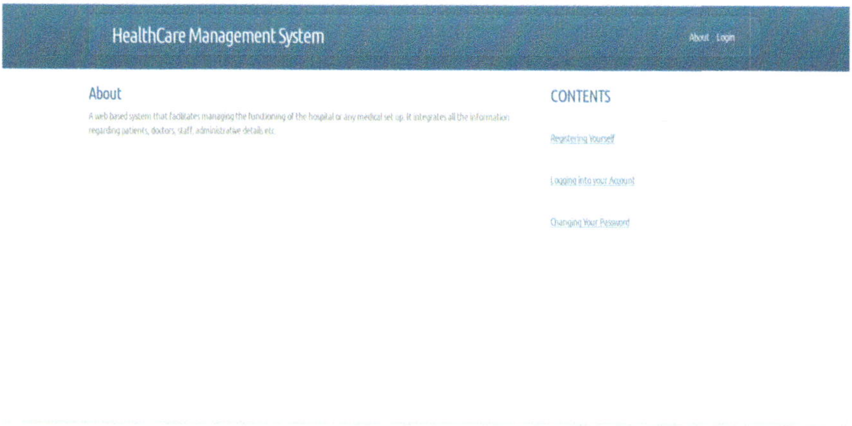

Fig. 10 System web portal

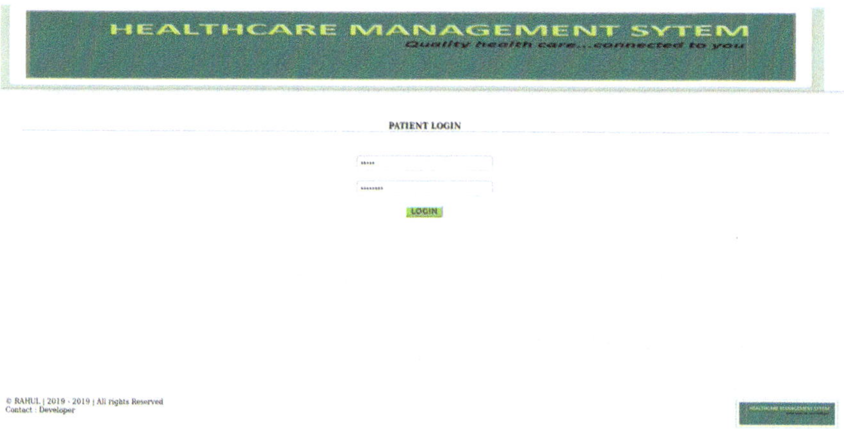

Fig. 11 Patient Login

The data from various sensors are uploaded in the phpMyAdmin server which is seen by the patient as well as a doctor which can be further used to analyze the health report (Fig. 17).

Fig. 12 Patient profile

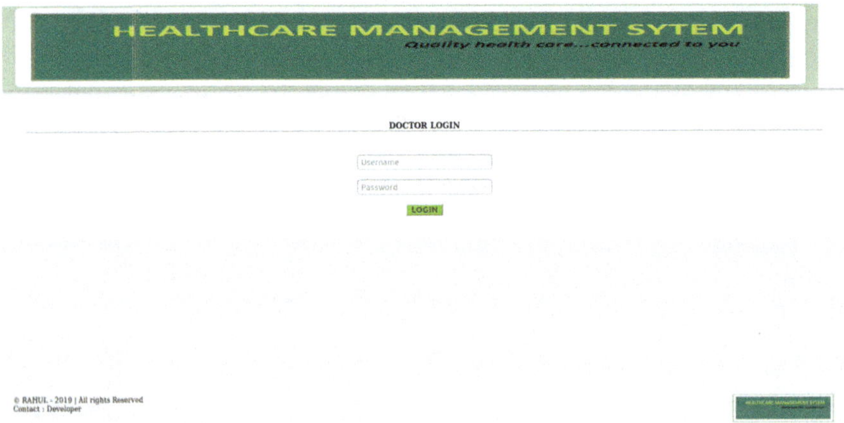

Fig. 13 Doctor Login

5 Conclusion

The discussed setup is a IoT-based system that can be used to keep a patient under observation. A patient, after getting discharged from hospital or clinic, can be kept under this observation. The system has three main concerns that is, temperature, body position and heart rate. These are measured with help of the fitted sensors. Finally, the measurements or data is sent to the local server, for analysis which is accessed by the experts and doctors. The portal also allows maintaining and tracking 24-h records of more than one patient. Which can be accessed from the web page by the doctors for more details. At present, we are accessing the health information on localhost,

Fig. 14 Doctor home page

Fig. 15 Patient registration

further, it can be extended over globally, after which a doctor will have the liberty to examine his patient from anywhere and anytime.

This proposed technique can also be integrated with the mobile by developing a smart phone app so that the model becomes more mobile and health information can be easily accessed at all times and at all locations.

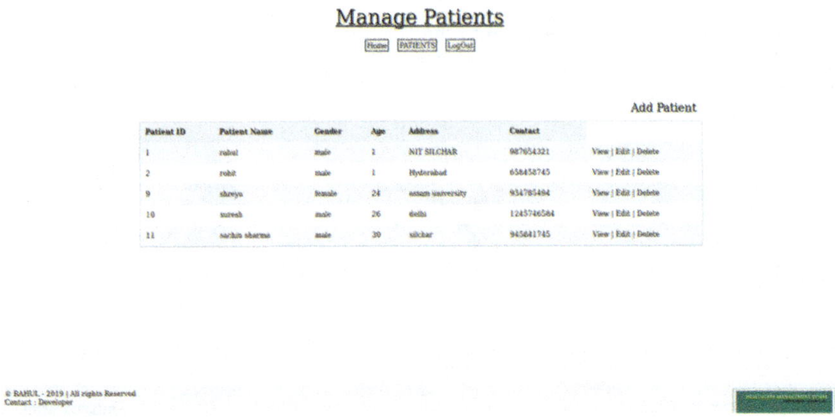

Fig. 16 Manage patient

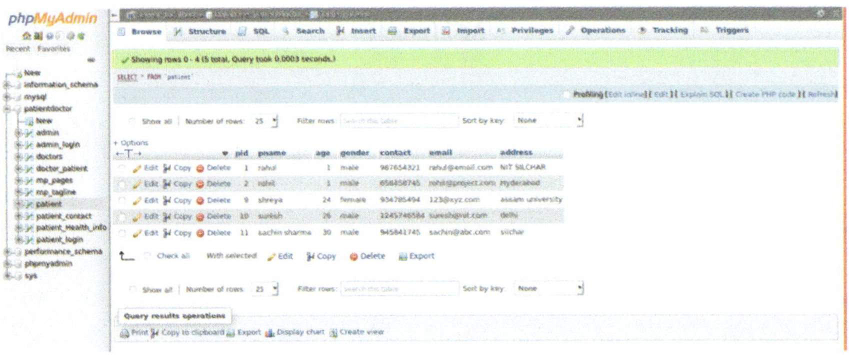

Fig. 17 Database server

References

1. Castillejo, P., Martinez, J.-F., Rodriguez-Molina, J., Cuerva, A.: Integration of wearable devices in a wireless sensor network for an E-health application. IEEE Wireless Commun. **20**(4), 38–49 (2013)
2. Budida, D.A.M., Mangrulkar, R.S.: Design and implementation of smart HealthCare system using IoT. In: International Conference on Innovations in Information, Embedded and Communication Systems (2017)
3. Sivagami, S., Revathy, D.: Smart health care system implemented using IoT. Int. J. Contemp. Res. Comput. Sci. Technol. **2** (2016)
4. Gupta, P., Agrawal, D., Chhabra, J., Dhir, P.K.: IoT based smart healthcare kit. In: International Conference on Computational Techniques in Information and Communication Technologies (2016)
5. Shukor, A.Z.H., Natrah, N.A.: Analysis of sitting posture recognition using pressure sensors. Int. J. Telecommun. Electron. Comput. Eng. **10** (2018)
6. Wenyao, X., Huang, M.-C., Amini, N., He, L., Sarrafzadeh, M.: eCushion: a textile pressure sensor array design and calibration for sitting posture analysis. IEEE Sens. J. **13**, 3926–3934 (2013)

7. Archit Sharma, K., Kumari, A., Singh, S.: A smart patient health monitoring system using Raspberry Pi 3. Int. J. Emerg. Technol. Innov. Res. (JETIR) **4** (2017)
8. Kamiya, K., Kudo, M., Nonaka, H., Toyama, J.: Sitting posture analysis by pressure sensors. In: International Conference on Pattern Recognition (2008)
9. Archip, A., Botezatu, N., Şerban, E., Herghelegiu, P.-C., Zală, A.: An IoT based system for remote patient monitoring. In: International Carpathian Control Conference (2016)
10. Huang, M., Gibson, I., Yang, R.: Smart chair for monitoring of sitting behavior. In: International Conference on Design and Technology (2017)
11. Sundaravadivel, P., Mohanty, S.P., Kougianos, E., Yanambaka, V.P., Ganapathiraju, M.K.: Smart-walk: an intelligent physiological monitoring system for smart families. In: International Conference on Consumer Electronics (2018)
12. Satija, U., Ramkumar, B., Sabarimalai Manikandan, M.: Real-time signal quality-aware ECG telemetry system for IoT-based health care monitoring. Internet Things J. (2017)

Data Privacy in IoT Network Using Blockchain Technology

Tamar Das and Shyamapada Mukherjee

1 Introduction

Blockchain is a decentralized and peer-to-peer distributed digital ledger which takes only similar types of information also called records and store that records in a series of blocks. The recorded order stored into the chain in a chronological way. When fresh blocks are appended to the chain with the help of consensus algorithm, it continues to expand. Here the blocks are connected together via a cryptographic mechanism. The blocks create a data chain altogether called asset, it may be tangible or intangible moves from place to place. These blocks give us the accurate timing and the transaction sequence. Each block is securely linked together to prevent any block from being changed or inserted between two other blocks. Blockchains are based on the concept of digital networks. In such networks, data transmission equates to transcript data from one position to another; such as, in the cryptocurrency world, this equates to transcript digital currencies from one person's digital wallet to some other person. The main difficulty is that the system must verify that coins are consumed one time and there will be nothing like double-spending problem. The typical approach is to engage a trust worthy middleman, for example, a central banking system, who acts as a trusted middleman between executing members and whose responsibility it is to maintain, preserve, and keep the data up to date. Concurrency control can be achieved by a central authority, and updates in the ledger if numerous users require to write in the digital ledger at the similar moment. Centrally control management system may not be always helpful or desirable in some cases, as it incurs intermediary expenses and forces internet users to place their faith in a third party to run the process smoothly [1]. Centralized process also offers a number of drawbacks because of single point of failure. This gives more endangered to both technological faults and hostile attacks [2]. Figure 1 represents a transaction process of bitcoin. A new transaction entered

T. Das · S. Mukherjee (✉)
National Institute of Technology Silchar, Silchar 788010, Assam, India
e-mail: shyama@cse.nits.ac.in

S. Mukherjee et al. (eds.), *Intelligent Systems for Social Good*,
Advanced Technologies and Societal Change,
https://doi.org/10.1007/978-981-19-0770-8_10

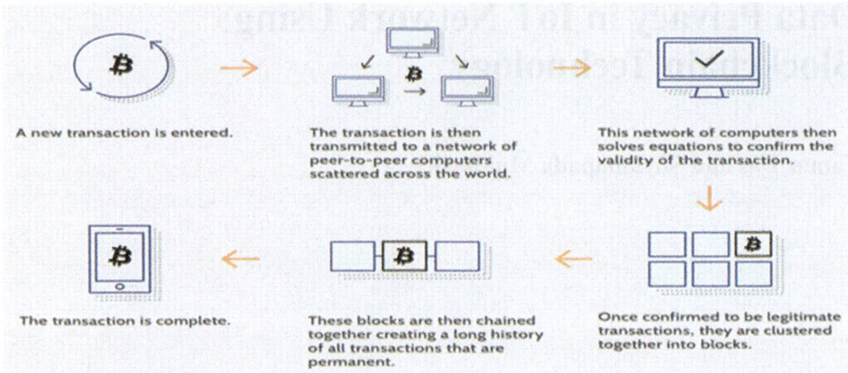

Fig. 1 Blockchain network records transactions on a digital ledger [3]

into the system then transmitted to a peer-to-peer network of computers scattered across the world then solving the cryptographic puzzle to confirm the validity of the transaction. Once the confirmation is completed, they are clustered together into the blocks. Then these blocks are chained together and create a long history of all the transactions and make it permanent. so ultimately the transactions are completed successfully.

The fundamental goal of blockchain technology is to eliminate the need for such kind of middleman and exchange them with a disperse network of computerized persons who collaborate to validate transactions and ensure the ledger's unity. Unlike central processes, each user of the blockchain technology owns a duplicate of the ledger. As a result, we can say that anyone who are available on the network can view the system's historical log and check its legitimacy, allowing for a high level of transparency. The challenge becomes finding an efficient solution to aggregate and synchronize many versions of the ledger if central administration is removed [4].

Figure 2 represents the characteristics of blockchain The integration of numerous core technologies example cryptographic hash function, digital signature which stands on asymmetric cryptography technique where public and private keys are present separately and distributed consensus technique allows blockchain to work in a decentralized context. So, we can say that blockchain has the potential to significantly reduce amounts and also increase efficiency [6]. According to market capitalization Bitcoin is the most well-known blockchain application nowadays, but this technology is not limited here, we can use it in a broad range of implementation beyond cryptocurrencies. Blockchain may be utilized in many financial sector-like virtual property, settlement, and also in digital payment system because it admits payments to be completed without any help of the central bank or a middleman. Furthermore, blockchain technology is emerging among the most auspicious technologies for the upcoming generation of interaction systems for the Internet, such as smart contracts technology, Internet of things, and Security services system. Blockchain technology offers enormous promise for the development of next-generation Internet services.

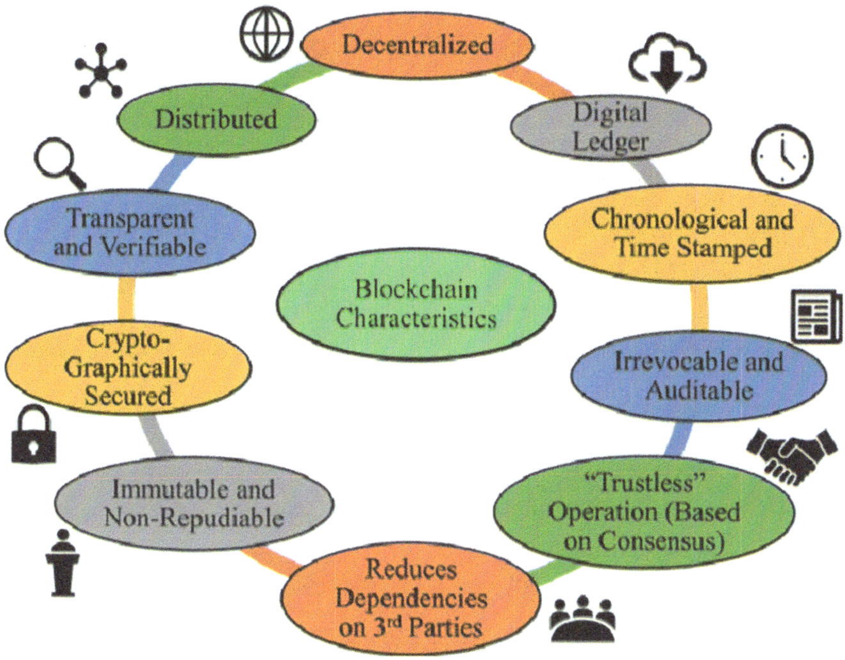

Fig. 2 Pivotal characteristics of blockchain [5]

This technology is also beset by the technical difficulties. Starting with, the scalability problem, it is a huge challenge for blockchain. The size of a Bitcoin blockchain is currently limited to 1 MB, and every fresh block, i.e., new block mining takes 10 min. As a result, we can say that the Bitcoin technology network is bounded to a limited number of transactions, i.e., only 7 transactions/second. On the other case, large size blocks require huge capacity space and also take longer for propagating via the network system. Since people desire to balance this large blockchain system, it slowly moves from decentralization to centralization system. After seeing these things, we can say that searching a balance in between block shape and security system has become a challenge. Second, it has been seen that the miners who mine the blocks can get reward more than their actual share money by using a selfish mining technique. So, blockchain branches might fall out frequently, obstruct blockchain outgrowth. As a result, some remedies require to put forward to fix this problem [6]. It's even possible to track a user's genuine IP address. Additionally, current consensus techniques such as proof of work (PoW) and proof of stake (PoS) have various major problems. PoW, for example, consumes lots of electricity, and the phenomena of the rich getting richer could show up, in case of the PoS consensus techniques. These types of issues must be solved in the development of blockchain system. Figure 3 represents the structure of the block in the blockchain system.

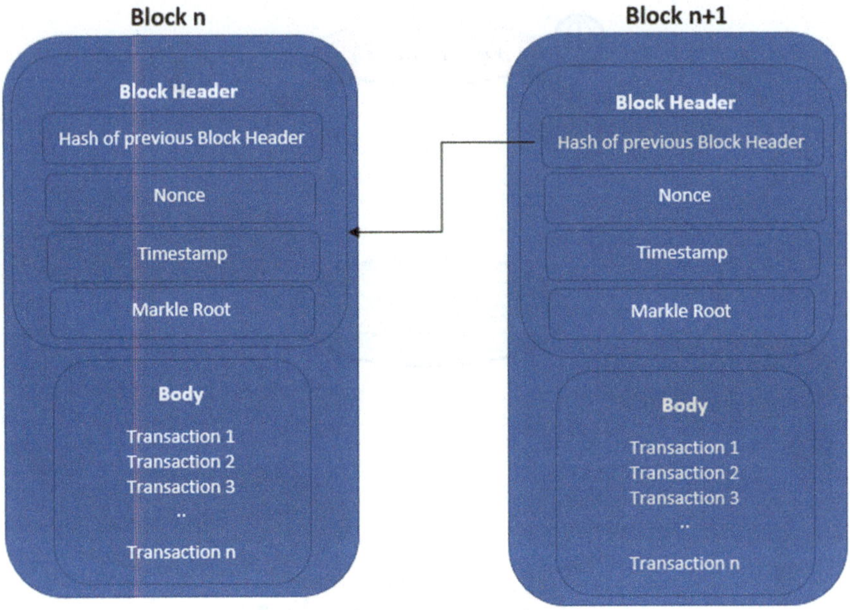

Fig. 3 Block structure [7]

The internet of things, or IoT, is a network of interconnected computing devices, digital and mechanical equipment, humans or animals, and objects that can perceive, gather, and transport data over the Internet without the need for human intervention. An IoT is created by web-enabled smart components that collect, send, and act on data from their neighbors using embedded systems example, processors, sensors, and communication hardware. By associate with an IoT gateway or other edge device, IoT devices can share sensor data that is sent to the cloud system for analysis purpose or maybe analyzed locally. Artificial intelligence (AI) and machine learning can also be helpful for IoT to make data collection processes simple and also dynamic.

2 History of Blockchain

In his 1982 dissertation "Computer Systems Established, Maintained, and Trusted by Mutually Suspicious Groups," cryptographer David Chaum suggested a Blockchain-like system for the first time. Stuart Haber and W. Scott Stornetta proposed it in 1991 as a way to design a system that would prevent document timestamps from being tampered with. Nick Szabo, a computer scientist, began working on 'bit gold,' a decentralized digital money, in 1998. Stefan Konst presents his theory of cryptographic protected chains, as well as implementation suggestions, in the year 2000. A white paper describing the model for a Blockchain was released in 2008 by the developer(s) operating under the pseudonym Satoshi Nakamoto. In 2009, he released the first Blockchain, which served as the public log for Bitcoin transactions. In 2014, the term "blockchain 2.0" was established in response to applications other than currency. Nick Szabo developed smart contracts in 1994, based on Blockchain. Smart Contracts are a set of executable codes that can run directly on top of Blockchain platforms. This technology system facilitates agreement between untrustworthy parties without the involvement of a third party. In contrast to Bitcoin, smart contracts are authorized in monetary exchanges.

3 Blockchain Architecture

Similar to a traditional public ledger, the blockchain is a series of blocks that carry a full list of transaction data. Some basic terminologies related to blockchain technology:

- Distributed Systems, this is a computing paradigm in which two or more nodes collaborate in a coordinated manner to achieve a common goal. The system's nodes are all capable of sending and receiving messages to and from one another.
- Asymmetric Key Encryption, this technology also known as the public-private key cryptosystem is used to construct identities on a Blockchain. The user creates two elements, a public key that is used to identify their transactions on the Blockchain system and a private key that is required to complete a transaction with the public key.
- Hash Values, similar to RPG (random password generator), this is an algorithm that is used to turn a validated transaction into a set of unique numbers and letters. Bitcoin generates hash numbers using the SHA256 algorithm. Bitcoin is a decentralized cryptocurrency system.
- Peer-to-Peer Networks (P2P) are a type of peer-to-peer network that allows computers connected in a network to connect directly to each other without the need for a central authority's reference, command, or routing, as well as the sharing of files, computational resources, and network bandwidth among those in the network.
- Merkle Trees are a type of data structure that is based on hashes. It resembles a tree structure, with each leaf node representing a hash of a data block and each non-leaf node representing a hash of its successors. It is mostly used for efficient data verification in distributed systems.

Blockchain is a cryptography secured hash function. It maps any sized data to a fixed sized. For example, H(p) = p % n, where p and n are integers and % is modular (remainder after division by n) operations. P can be of any arbitrary length, but H(p) is within the range [0, n − 1]. P is called the message and H(p) is called the message digest. Cryptographically secured means given a p, we can compute H(p), but given a H(p), no deterministic algorithm can compute p. For two different values suppose p1 and p2, H(p1) and H(p2) should be different. There are various cryptographic hash functions available for example, MD5, SHA256, etc. Bitcoin cryptocurrency

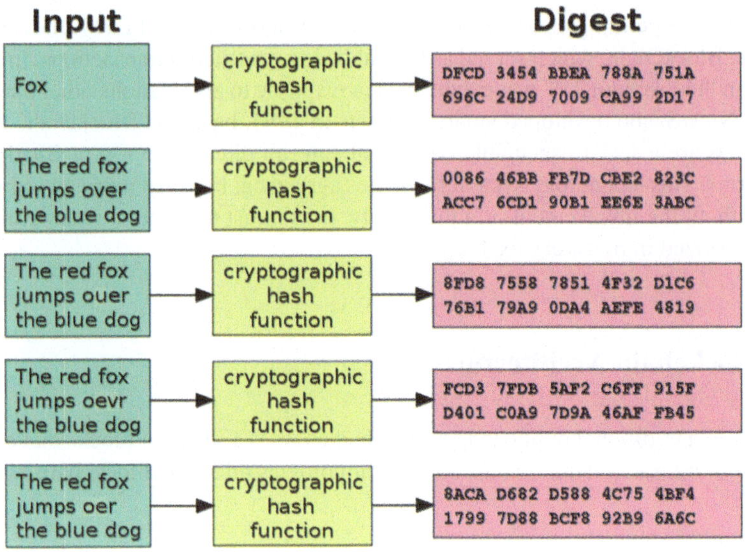

Fig. 4 Cryptographic hash functions and collisions [8]

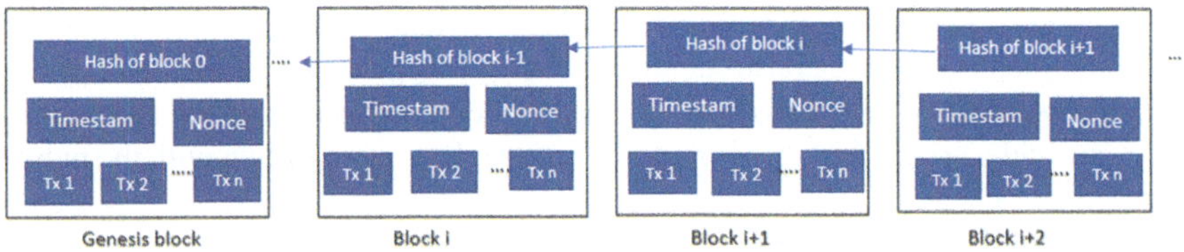

Fig. 5 An example of blockchain which consists of a continuous sequence of blocks [6]

uses SHA256 function for generating hash value. If we want to little change in the data set it will affect the results huge. This is called Avalanche effect.

Figure 4 represents for different input messages the output are expected to produce like different hash values, i.e., called message digest.

The initial block of a blockchain series is called a genesis block. This type of block doesn't have any parent block. Parent block is a block where every block has a connection to the other block before it. Figure 5 represents a continuous sequence of blocks. How blocks are connected together via a chain system.

3.1 Block

Mainly the block header and the block body make up a block. Particularly the head of the block includes various things for example:

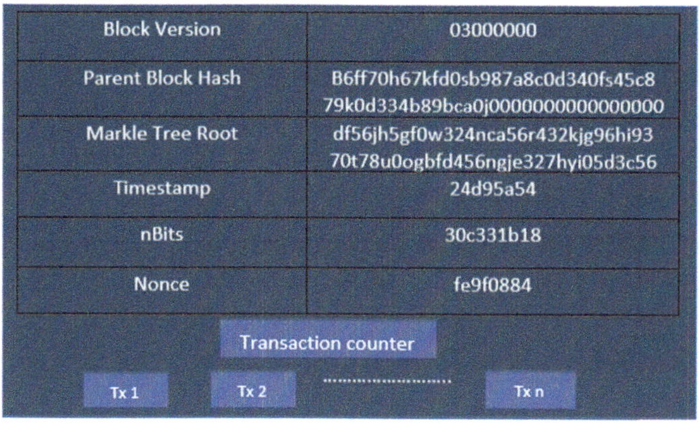

Fig. 6 Block structure [6]

- Block version: It is actually helpful for marking which set of blocks validation criteria that is going to be used.
- Root block hash: It is the value of hash of 256 bits that links to the preceding block.
- Markle tree: It is a tree that gives us the total of all the transactions that happen during the transaction time in the block's hash value.
- Timestamp: Every block in the blockchain system contains a timestamp which is called Unix time timestamp. They not only provide variance for the blockchain block hash, but also make manipulating the block chain more complex for an opponent.
- nBits: a concise representation of the current hashing target.
- Nonce: It is a 32-bits field that normally starts with zero(0) and rises with each hash computation. Figure 6 represents the details of a block.

3.2 *Digital Signature*

A numerical approach to present the legitimacy of digital messages or record is known as a digital signature. A genuine digital signature permits a receiver cause to faith that the message was set up by the claimed sender (authentication), and the sender later cannot disagree about it (non-repudiation), and that the message was not tampered with while in transit (integrity). Asymmetric cryptography is used in digital signatures. Asymmetric cryptography is commonly acknowledged as public key cryptography (PKI). PKI is used to encrypt and decrypt data using public and private keys. One of the keys in the couple, acknowledged as the public key, can be shared with anybody. The other key in the couple, acknowledged as the private key, is kept hidden. Any of the keys we can use for encryption a message; the decryption key is the inverse of the one used to encrypt the message. The transactions are digitally

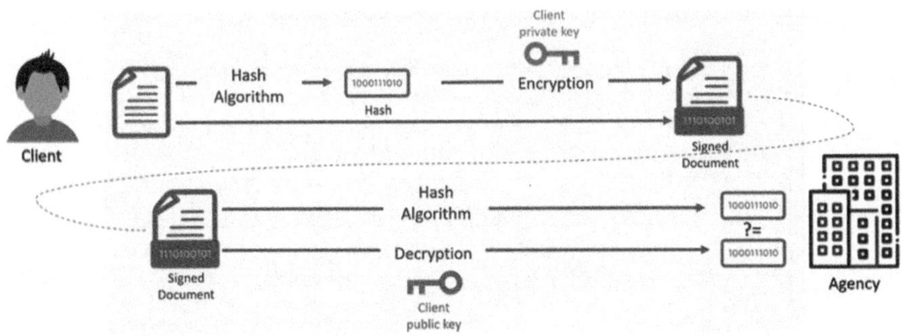

Fig. 7 Digital signature used in blockchain [6]

signed with the help of the private key. These signed transactions are dispersed over the network and usable using the public keys that are clear to each one in the networking system. The signed phase and the verification phase are the two phases which are involved in a conventional digital signature. Here we are going to show an example regarding these systems, Alice, a user when wishes to sign a transaction, she first produces a hash value coming from the transaction system. After that she encrypts this hash function result with her private key and bypass the encrypted hash together with the real-data value to another user Bob. Bob confirms the received transaction by matching the decrypted hash value (using Alice's public key) and the hash function value coming from the received data using the similar hash function as Alice's done. The elliptic curve digital signature algorithm (ECDSA) is a common digital signature algorithm used in blockchains (Johnson et al. 2001).

Figure 7 represents how a digital signature works in block system.

3.3 *Blockchain Key Characteristics*

Decentralization: Every transaction in conventional centralized transaction systems should be validated by a central trusted authority (for example, central bank), deriving in price and production bottlenecks at the central servers. Blockchain does not need the help of a third party, unlike the centralized approach. Blockchain uses consensus methods for keeping data compatible over a distributed network [9]. Persistency: If the transactions are valid it would be accepted quickly, and if it is invalid it would not be accepted by the honest miners and called as an invalid transaction. Once a transaction is incorporated in the blockchain, it is nearly hard to remove or rollback the transaction. Blocks containing wrong transactions might be recognized right away. Anonymity: In this approach every user who are interacts with the blockchain using a unique address, conceals their true identity. Because of the inherent constraint of blockchain, it will not give the surety about the absolute privacy protection. Auditability: The Unspent Transaction Output (UTXO) model [9] is used to store record about participants amount on the Blockchain network. Any transaction must reference some formerly unused/unspent transactions. The state of those referred unspent transactions is modified from unspent to spent when the present transaction is listed in the blockchain network. So, tracking the transactions would be very easy and also verified.

3.4 *Taxonomy of Blockchain*

Depending on the desired function and use case, a blockchain network or system might follow variousregulations and system topologies. Network users and validators are the most common components of blockchain systems. User nodes have the ability to send and receive transactions, as well as keep a copy of the ledger. Validators are responsible for authorizing ledger alterations and obtaining network consensus on the ledger's valid state, moreover having read access privileges. Limited or universal access rights and validations rights may apply depending on the system design. There are mainly three types of blockchain systems now in use: public blockchain, private blockchain, and consortium blockchain [10]. A public blockchain system is open to any Internet user. Private block chains, on the other hand, limit access to only those who have been granted permission. Since each member of the network can contribute to transaction validation, permissionless ledgers are fully dispense and censorship-resistant. With permissioned ledgers, on the other hand, only a few validator nodes have the right access to the blockchain, allowing them to edit it. Users and validators are fully not known to one another in case of public and permission less ledgers. On the other hand, users' identities are known through private and missioned ledgers, akin to know-your-customer practices (KYC). Consortium blockchain also known as hybrids blockchain is the combination of public and private blockchains [6]. In case of consensus process both consortium and private blockchain are permissioned and public blockchain is permissionless. In case of centralized system public blockchain is not centralized, consortium is partial, and private blockchain is totally centralized. Efficiency is high in both the cases of consortium and private blockchain but in case of public it is low. In case of immutability, public blockchain is nearly impossible to tamper but in the other two blockchain systems it could be tampered. Some applications that are used blockchain concept nowadays

- Bitcoin: Bitcoin is the world's first cryptocurrency and also the largest by market capitalization. In 2008, the first Blockchain was conceptualized by Satoshi Nakamoto, an anonymous person by his publishes "Bitcoin: A Peer-to-Peer Electronic Cash System" [11]. In 2009, it was first effective to the people. It is mainly used SHA256 algorithm to "hash" data into a 256-bit number. Every member of the bitcoin system has taken a digital wallet, in which all the coins are stored. This wallet can be accessed only by user's private key. It uses a proof of work consensus algorithm.

- Ethereum: Ethereum is a decentralized and open-source Blockchain technology with the functionality of smart contract system. For Ethereum, Ether is the native cryptocurrency system. After Bitcoin cryptocurrency system, it is the second-largest market capitalization. It is not only bounded into cryptocurrency rather than that we can use it in various fields.
- Hyperledger fabric: Hyperledger Fabric is designed to serve as a basis for building applications and solutions with a modular architecture. It gives permission to the component, for example, consensus and membership services to be plug-and-play. Its modular and adaptable architecture caters to a wide range of industry applications. It takes a novel method to consensus that allows for scalability while maintaining anonymity.

4 Consensus Algorithm

Blockchain is a dispersive and decentralized network system that allows immutability, privacy, security, and transparency. In spite of the shortage of a central authority to validate and confirm transactions, the Blockchain considers each of the transaction systems to be entirely threat free and validated. All these things are possible because of the consensus algorithm and it is a crucial factor of any Blockchain network. It is an algorithm via which all the members of the networking system come to the same agreement about the current state of the distributed ledger. Now we'll look at a few different consensus algorithms and how they function:

- Proof of Work (PoW): This algorithm is used for selecting a miner, who help for creating next block in the blockchain. This strategy was used by satoshi nakamoto, an anonymous person for his bitcoin cryptocurrency in 2008. It is working by computing the hash values and validating the transactions till the hash function value carries a certain number of trailing zeros. A nonce is an integer that creates a hash function value with a fixed number of trailing zeros. In the hash function mechanism, a nonce is a random number that creates the required number of trailing zeros. It is a consensus algorithm for permissionless public ledgers that makes use of the computational resources of the node's systems. The bitcoin mining is the cryptographical puzzle to find a random number that produces hashes with a certain number of foremost zeros. The public and private keys issued to each user are used to validate and sign each transaction. In every 10 min, a Bitcoin block is formed, but in case of Ethereum it takes 17 s to create a block. It is a resource-intensive and power-hungry protocol. In comparison to more efficient protocols, this approach wastes a significant amount of energy and electricity to solve cryptographic puzzles. Figure 8 represents how a longer chain will be selected and the shorter chain discarded from the blockchain system.
- Proof of Stake: It is the top popular substitute for PoW. Ethereum has changed from PoW technique to PoS technique. Here stake holder needs to spend costly gear to solve a complex puzzle. Validators mean that the user spends in the system's cryp-

tocurrency using lock up few of their coins as like stake in the structure of consensus process. After that the transaction blocks will be affirmed by every validator. It is an energy saving strategy. Rather than needing participants in the networking system to search a nonce in an infinite storage, POS needs users to show the ownership of the quantity of cryptocurrency as it is assumed that those with more currencies will be less chances to invade the network system. The choice based on available wallet balance is highly unjustifiable since the individual wealthy/affluent person is certain to be prominent in the networking system. Subsequently, variousresults are presented using a mix of stake size to determine which one to produce the upcoming block. Blackcoin in particularly exploits randomization to forecast the upcoming generator. It employs a formula that considers the lowest hash value as well as the stake size. Coin age-based selection is favored by Peercoin. Older and larger set of coins have a higher probability of mining the upcoming block in Peercoin. When we analyze between PoW and PoS we can see that PoS uses less energy as compared to PoW and also more efficient. Unfortunately, because the value of mining is too low, attacks may occur as a consequence.

- Practical byzantine fault tolerance (PBFT): Practical Byzantine fault tolerance (PBFT) is a transcript algorithm that allows byzantine faults to be tolerated. Because PBFT can handle up to 1/3 malicious byzantine replicas, Hyperledger Fabric (Hyperledger 2015) uses it as its consensus method. In each and every round, a new block is destined. The initial would be chosen based on some rules and regulation in each round. Pre-prepared, prepared, and commit are the three phases of the procedure [6]. Hashing techniques are not applicable here in PBFT. Each node in PBFT is required to query other nodes, whereas Stellar Consensus Protocol (SCP), A Byzantine agreement protocol, allows participants to select which groups of other users to trust. SCP Antshares [12] established their delegated byzantine fault tolerance (dBFT) based on PBFT. Instead of all nodes, some professional nodes are voted to record transactions in dBFT.
- Proof of Authority (PoAu) To generate a block with PoAu, one or more members must be granted special authorization to make changes to the blockchain. One person with a specific key, for example, could be in charge of generating all the blocks. PoAu is essentially a modified PoS algorithm in which the stake for validators is their own identity. Members of the network place their trust in authorized nodes, and a block is accepted if it is signed by the majority of authorized nodes. Voting can be used to add new validators to the system [13].
- Proof of Space Proof of Space, often known as PoSpace, is a network consensus protocol that works in a similar way to Proof of Work. Proof of Space validates transactions using disc storage rather than computing resources. It consumes disc space and rewards miners who have the largest disc space allotted to a block. This data structure is used to solve the pebbling game and is implemented using hard-to-pebble graphs. Pebbling vertices in a graph is only possible if all of the parent vertices have been pebbled [14].

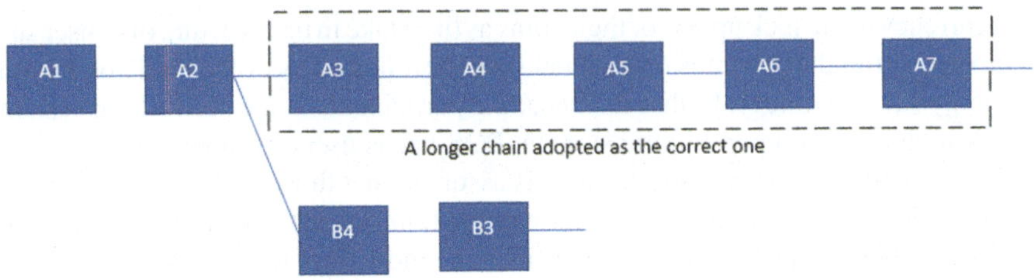

Fig. 8 A scenario of blockchain branches (the longer chain would be accepted as the main chain and the shorter one would be discarded)

5 Application of Blockchain

There are various types of applications available for blockchain technology. We can apply this technology in Financial services sector, healthcare system, government sector, travel and hospitality management system, retail and consumer packaged goods, e-Notary service, identity management, IOT device management and security service, eVoting system, etc. Now we will discuss few of them for example

- Finance sector: Blockchain technology has already been used in a variety of innovative ways in the financial services industry. By offering an automated trade lifecycle where all users have access to the same facts about a transaction, blockchain technology simplifies and streamlines the entire process connected with asset management and payments. This eliminates the need for brokers or intermediaries, ensuring transparency and efficient data handling.
- Healthcare system: Blockchain has the potential to transform the healthcare industry by improving the privacy, security, and interoperability of patient data. It has the potential to address many of the industry's interoperability issues and enable secure data sharing among the many companies and persons engaged in the process. It removes third-party influence while also avoiding overhead costs. Healthcare records can be kept in distributed databases using Blockchains by encrypting them and using digital signatures to assure privacy and validity [15].
- Government sector: The power of blockchain technology to transform government operations and services is undeniable. It has the potential to significantly improve data transactional difficulties in the government sector, which now operates in siloes. The proper linking and sharing of data with blockchain allow for better data management across departments. It increases transparency and makes it easier to monitor and audit transactions.

Figure 9 represents the various applications of Blockchain in different areas.

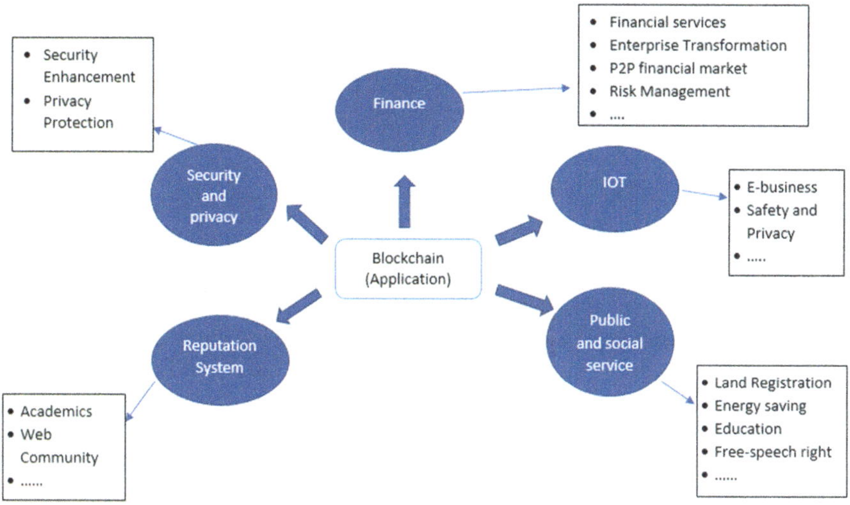

Fig. 9 Representative application domains of blockchain [6]

6 Challenges in Blockchain

- Performance and scalability, it is coming under technological challenges where decentralized architecture works slower than the traditional systems. Performance affected because of calculations connected with encryption and decryption and hashing function at each node. Scalability is affected by the factors, for example, architecture, network bandwidth, configuration of block, block size, variable requirements for processing power, file system, Consensus algorithm, Transaction validation mechanism, etc [25].
- Storage, Blockchain technology allowed to append data storage mechanism. After storing data into the Blockchain it will not be changed, it becomes continuous and also available at every node in the networking system. In this case it urges more resources and may create an issue when the chain grows.
- Privacy, Blockchain information are stored in each block in the networking system and so the privacy is not a hidden feature that Blockchain conventionally provides us. Here the data need to be captured in such a way that the privacy security of a single person is not leaked to all.
- Regulations, the state of regulations and willingness for Blockchain applications are still suspicious. It is principally related to the privacy system of the information allocation through the Blockchain technology which may be user identity documents or health records. The adoption would be smooth and quicker when the regulations are well defined.
- Security, use of current Certificate Authorities—each entity with the node and the participants in the Blockchain network have a public key, a private key, and

also a certificate. The choice of certificate authority would rely on the nature of application that is targeted mainly.
- High development cost, the creation of block in the Blockchain technology takes high cost in terms of power consumption and other things.

6.1 Risk with Blockchain Technology

- 51% attacks, in case of Mining it requires a huge amount of computing power, especially for large-scale public Blockchains. It occurs when a single cryptocurrency miner or group of miner's gains administers in excess of 50% of a network's Blockchain, which means one or more miners takes administers in excess of 50% of a network's mining authority then computing power or hash rate. If it is successful, the miners responsible essentially control all the network and certain transactions occur within it.
- Proof of Stake vulnerabilities, this attack is also known as Bribe Attack. Here the attacker carries out a consuming transaction that he wishes to overturn later. Instantly after the transaction happen, the attacker tries to start to make another chain which is based on the block prior to the one presenting the transaction. The attacker creates an another chain in totally confidential. Latterly the transaction acquires the important number of affirmations and the attacker's digital chain is lengthy than the original chain, the attacker brings out this entirely. The attacker's chain is acknowledged as the fresh substantial Blockchain and the transaction is switched.
- Double spending, there is a danger that a member with, for instance, one bitcoin can spend it twice and falsely receive goods to the value of two bitcoins before one of the providers of goods or services realizes that the money has already been spent. The most serious threat for double-spending case is 51% attack problem. This problem arises when the network user administers larger than 50% of the computing power for preserving the distributed ledgers of the cryptocurrency system. If these members administer the Blockchain network system they can be exchange bitcoins to their personal digital wallet various times by overturn the Blockchain ledger.

7 Introduction of IoT

The internet of things, or IoT, is a linked network of computing devices, electronic and mechanical equipment, humans or animals, and objects that can perceive, gather, and transport data over the Internet without the need for human intervention. It is invented by web-technology-based smart devices that arrange, send, and react on data from their nearby embedded systems such as processors, sensors, and conveying hardware. IoT devices can communicate sensor information by connecting to an IoT platform or other edge device, which can then be forwarded to the cloud for analysis or examined locally. Artificial intelligence (AI) technique and machine learning mechanism these two techniques also used by IoT to assemble data for transaction easier and also more dynamic.

8 Architecture of IOT

There are mainly four stages available. Figure 10 represents these four stages.

- Stage 1: Sensors and Actuators: Sensors and actuators are the components for sending, receiving, and processing the data all over the network system. These components are merged with cable wire or may be wireless systems. RFID, GPS, electrochemical, gyroscope, and other technologies are also included in these technologies. Local Area Network or Personal Area Network can be helpful to connect sensors and actuators altogether [16].
- Stage 2: Data Acquisition System: This system is mainly used for collecting new/fresh value from sensor side and converts these values from analogue to digital format. Then combine and prepare the data before transmitting it to the upcoming level of processing with the help of an Internet gateway, wireless WANs (such as Wi-Fi or Cellular), or wired WANs [17].
- Stage 3: Preprocessing units: Data analytics: After the IoT data has been digitized and aggregated, it will need to be processed to reduce the data amount even further before being sent to a data center or cloud. As part of the pre-processing, the edge device may execute some analytics [16].
- Stage 4: Smart application and management: Data center or cloud: The Management Services group includes the Data Center or Cloud, which processes data through analytics, device management, and security controls. Apart from security controls and device management, the cloud sends information to end-user applications including retail service, healthcare system, emergency case, environment issues, energy resources, and among others also [17].

9 Application of IoT

We can use IoT technology everywhere. For example in agriculture sector produce good quality crops using smart farming technology and also check the soil quality, in smart home and building system it can helpful, then mobile devices, energy sector, Health sector, Industrial sector, security and services sector, and in wireless sensor network also IoT plays a vital role. Here in Fig. 11 showing that we can use this technology in various fields. It is working on only IoT-enabled components and can be created to be specific to almost every industry.

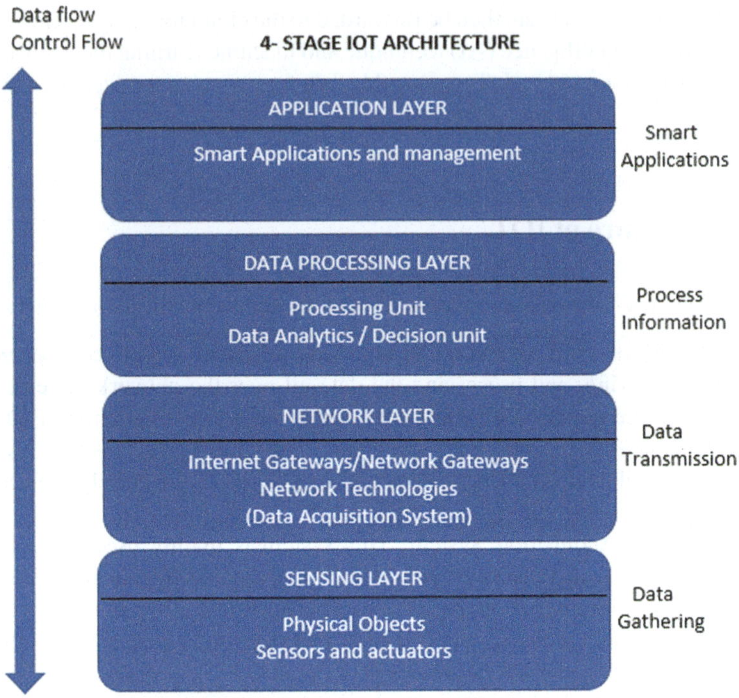

Fig. 10 IoT architecture

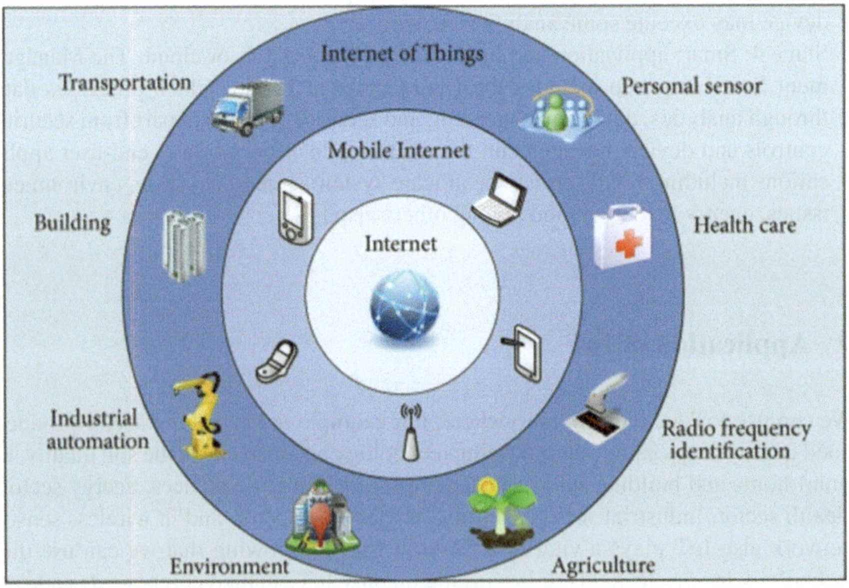

Fig. 11 IoT in various field [18]

10 Challenges of IoT

There are various challenges facing by IoT systems. Many industries have been affected by the Internet of Things, for example, healthcare sector, information technology system, data analytics, agriculture sector, etc. The primary focus is on maintaining privacy system, as it is the fundamental cause of other issues. To avoid IoT from being impeded, a combined effort from the government side, civil society, and the private sector would be crucial.

- Scalability: When billions of internet-enabled devices are linked together in a massive network, massive amounts of data must be processed. Scalability is required for the system that saves and analyzes data from these IoT devices. Everyday things are now connected to each other over the Internet in the era of IoT evolution. For the interpretation of valuable data generated from these devices, big data analytics and cloud storage are required [19].
- Data privacy, confidentiality, and integrity: Since IoT data goes via numerous hops in a networking system, it needs to use an encryption mechanism so that we can protect data confidentiality. The data which is saved on a device is exposed to privacy violations by accommodating nodes in an IoT network system because of a diversified integration of services, devices, and networks. Since IoT devices are vulnerable to attacks, an attacker could compromise data integrity by changing stored data for malevolent objectives [20].
- Lack of government support: Government and regulatory authorities, such as the FDA, should create regulations by establishing a standard committee for devices and person's safety and security.
- Single points of failure (SPOF): Nonstop growth of heterogeneous network systems for IoT-based architecture may expose a huge number of SPOF, resulting in deterioration of IoT-based services. It necessitates the creation of a tamper-proof environment for a huge number of IoT devices, as well as alternate ways for fault-tolerant network deployment.
- Energy efficiency: IoT devices are often resource constrained, with minimal power consumption and limited storage capability. By flooding the network system and draining IoT resources through duplicate or fraudulent service requests attacks on IoT systems may result in an increasing order in the field of energy usage.

11 Application of Blockchain in IOT

One of the most that are incomparable potentiality, information, and communication technologies, the Internet of Things (IoT) is elevating lately. IoT is intended to connect things to the internet and offer consumers with variousservices [21]. Logistics control with Radio-Frequency Identification (RFID) technology, smart home system, e-health facility system, smart grids system, and so on are some of the most common killer applications of IoT. Here are some of Blockchain Enterprise use cases that show how Internet of Things and Blockchain combinedly may create a big impact in a variety of industries such as Automotive, Smart Home system, Sharing Economy system, Pharmacy, Agriculture sector, Water Management, Supply Chain, and Logistics system. Here we will discuss some of them.

- Smart Homes System: Smart IoT-enabled devices are becoming increasingly important in our daily lives. The Internet of Things blockchain permits home security systems to be controlled remotely from a smart device like smartphone. On the other hand, in the typical centralized way to exchanging data created by Internet of Things devices have shortage of security requirements and data ownership. Blockchain addresses security concerns and eliminates infrastructure that is centralized which will help to create smart home system. Telstra, an Australian telecommunications and media business, has adopted blockchain and biometric security to make sure that data collected from smart devices cannot be tampered with. For increased security, sensitive user data is maintained on the blockchain, such as voice recognition, face recognition, biometrics. When the information is saved on the blockchain system, it can't be changed, and only the right people have permission to access it [4, 22].
- Logistics and Supply Chain: Multiple parties are involved in a supply chain network, this is why delivery delays are one of the most significant difficulties in the supply chain and logistics business. This one is the place where both Blockchain and IoT come into play. IoT-enabled components allow employees to trace shipments at each and every phase and Blockchain ensures that the total process is transparent. Various IoT sensors like motion sensors, GPS sensor, temperature sensors, and so on can provide information regarding the position of a shipment. After that the data will be posted on the Blockchain system for clarity purpose; after the value is stored on the Blockchain, every supply chain holder named in the Smart Contracts and have real-time permission to access data value. Both Blockchain and IoT can help together to improve the supply chain network's traceability and efficiency [4].
- Pharmacy Industry: The growing cases of forged drugs are one of the pharmaceutical industry's biggest challenges nowadays. The pharmacy sector can now fight this problem because of Blockchain IoT technology. In Blockchain IoT all the stakeholders which are involved in the medicine production system may be held accountable and modernize the blockchain system with some important data of the present time. Mediledger is a Blockchain IoT application that can trace the legitimate change of prescription of drug owner [22]. Represents the game-changer concept in the pharmacy industry
- Agriculture industry: Rising the production of food for the increasing demography is a big challenging nowadays. So to overcome from this situation we can use the technology system of blockchain and IoT altogether. Blockchain and IoT both have the potential to completely change the food production system. Here using these technologies, we can transform food industry to farming and then from farming to grocery and then finally to house. So here we need to Install IoT sensors into the farming place and transfer the information straight to the blockchain

system. In that way we can assist to improve the food supply system chain. Pavo is a blockchain IoT system that delivers transparency which are not paralleled to farmers by presenting a fresh and smart farming technique [22].

11.1 Some of the Companies Who Are Using IoT Blockchain

There are various companies available in the market who are using IOT Blockchain technology system together for the society example:

- Helium: It is the first decentralized machine network in the world. The company uses blockchain to connect low-power IoT gadgets to the Internet (such as routers and microchips). It is basically a San Francisco, California company [23, 24].
- Chronicled: It provides an end-to-end supply chain solution by combining blockchain and IoT products. The pharmaceutical and food supply businesses are the main focus of this company. It is also a San Francisco, California-based company [23, 24].
- Arctouch: It creates blockchain-based software for voice assistants, wearables, and smart TVs, among other smart and connected devices. It is also a San Francisco, California-based company [23, 24].
- Filament: It creates blockchain-based hardware and software products that connect seamlessly with IoT devices. It is basically a Reno, Nevada-based company.
- Hypr: It secures connected ATMs, vehicles, locks, and homes using decentralized networks. It is a centralized databases system that can store millions of passwords. It is a New York based company.

12 Future Work

- Security and scalability issues: As we know that IoT security system is a serious issue that has impeded its widespread adoption. IoT devices are often vulnerable to security flaws, create them a main target for Distributed Denial of Service (DDoS) attacks. Different DDoS assaults have wreaked havoc on businesses and personal in recent years. Additional problem with recent time IoT network system is scalability issues. Current centralized methods to authenticate, authorize, and link diverse nodes in a network system will become a bottleneck when the number of devices connected via an IoT network grows. So, in these field anyone can work with the help of Blockchain. Because it has the capability to help in security and as well as scalability.
- Smart contract: It is another domain where the people can work. As we know that, one of the biggest challenges for IoT is that it currently relies heavily on centralized platforms. Hackers may be able to access sensitive data as a result of this. This is a problem that blockchain technology has the ability to solve. Blockchain is

a decentralized digital ledger of transactions that preserves information in such a way that it prevents the data from being hacked or altered. It accomplishes this by replicating transactions and spreading them throughout the network to "nodes." Smart contracts, as defined by blockchain, would be able to be performed independently and decentralized way, resulting in a web of connected devices that provides people authority over their own data.

13 Conclusion

In this study we discussed various applications and challenges of Blockchain and IoT and also study how blockchain helps IoT system in various fields, for example, agriculture, health sector, etc. So the combination of blockchains and IoT can be strong in various fields. Blockchains provide us with resilient, distributed peer-to-peer systems and the capability to interact with peers in a trustless and auditable manner. Both IoT and blockchains are still in their infancy, but they both hold the promise of making machine-to-machine communication a breeze. Companies are currently working to combine the two technological powerhouses. IoT and blockchain technology, when combined, will allow various businesses to prosper by allowing them to easily monitor, track, and secure data. We think that the regular integration of blockchains in the IoT sector will induce significant changes across various industries and come up with new business models. We also discussed few possible future works. We are planning to take an in-depth investigation on Smart contract in future.

References

1. Akins, B.W., Chapman, J.L., Gordon, J.M.: A Whole New World: Income Tax Considerations of the Bitcoin Economy (2013)
2. Armbrust, M., Fox, A., Griffith, R., Joseph, A.D., Katz, R., Konwinski, A., Lee, G., Patterson, D., Rabkin, A., Stoica, I., Zaharia, M.: A view of cloud computing. Commun. ACM **53**(4), 50–58 (2010)
3. https://www.investopedia.com/terms/b/blockchain.asp
4. https://www.upgrad.com/blog/blockchain-in-iot-examples-applications/
5. Puthal, D., Malik, N., Mohanty, S.P., Kougianos, E., Yang, C.: The blockchain as a decentralized security framework. IEEE Consum. Electron. Mag. **7**(2), 18–21 (2018)
6. Zheng, Z., Xie, S., Dai, H.-N., Chen, X., Wang, H.: Blockchain challenges and opportunities blockchain challenges and opportunities: a sey. Int. J. Web Grid Serv. (IJWGS) **14**(4) (2018)
7. National strategy on blockchain, Government of India Ministry of Electronics and Information Technology (MeitY) (2021)
8. https://wizardforcel.gitbooks.io/practical-cryptography-for-developers-book/content/cryptographic-hash-functions/crypto-hashes-and-collisions.html
9. Zheng, Z., Xie, S., Dai, H., Chen, X., Wang, H.: An overview of blockchain technology: architecture, consensus, and future trends. IEEE Int. Congr. Big Data (BigData Congress) **2017**, 557–564 (2017)
10. Buterin, V.: On Public and Private Blockchains. https://blog.ethereum.org/2015/08/07/onpublic-and-private-blockchains/ (2015)

11. Nakamoto, S.: Bitcoin: Apeer-to-peer electronic cash system. https://bitcoin.org/bitcoin.pdf. Accessed 15 July 2021 (2008)
12. Antshares.: Antshares Digital Assets for Everyone. https://www.antshares.org (2016)
13. Eurelectric, Eurelectric launches expert discussion platform on blockchain. http://www.eurelectric.org/news/2017/eurelectric-launches-expert-discussion-platform-on-blockchain/. Accessed 24 Aug 2021 (2017)
14. https://www.section.io/engineering-education/blockchain-consensus-protocols/
15. https://www.happiestminds.com/Insights/blockchain/
16. https://www.digi.com/blog/post/the-4-stages-of-iot-architecture
17. https://www.javatpoint.com/iot-architecture-models
18. https://www.scientechworld.com/blog/telephone-network-computer-network-mobile-network-network-things-internet-things-iot/
19. https://www.geeksforgeeks.org/challenges-in-world-of-iot/
20. Khan, M.A., Salah, K.: IoT security: review, blockchain solutions, and open challenges. Sci. Direct (2018)
21. Atzori, L., Iera, A., Morabito, G.: The internet of things: a survey. Comput. Netw. **54**(15), 2787–2805 (2010)
22. https://www.leewayhertz.com/blockchain-iot-use-cases-real-world-products/
23. ibm (2016) IBM Blockchain. http://www.ibm.com/blockchain/
24. https://builtin.com/blockchain/blockchain-iot-examples
25. Mattila, J., Seppala, T., Naucler, C., Stahl, R., Tikkanen, M., Badenlid, A., et al.: Industrial blockchain platforms: an exercise in use case development in the energy industry. https://www.etla.fi/julkaisut/industrial-blockchain-platforms-an-exercise-in-use-case-development-in-the-energy-industry. Accessed 10 Aug 2021 (2016)

Interval-Valued Fuzzy Parameterized Multi Fuzzy N-soft Set in Decision-Making

Ajoy Kanti Das and Carlos Granados

1 Introduction

The SST [44] is extremely useful in a variety of situations. Molodtsov [44] developed the basic results of SST and successfully applied it to a variety of fields, including the smoothness of operators, operations analysis, Riemann integration, theory of game, probability, and so on. Later, Maji et al. [38] presented several new SST concepts, such as subset, complements, union, and intersection, as well as their implementations in DMPs. Ali et al. [16] identified some new operations on SST and demonstrated that De Morgan's laws apply to these new operations in SST. To solve the DMPs, Maji et al. [39] used SST for the first time. Recently, several authors later looked into the more broad properties and applications of SST [9, 11, 31, 32].

Many academics are interested in hybrid models, as seen by the aforementioned references. Many hybridization options for the recently created NSSs [31] model. This model's primary role is to broaden the scope of SST applications, which deal with qualities that resemble the universe of discourse. Because many real-world examples have insisted on their applicability, this paradigm constitutes a practical expansion of SST (see [11, 31, 32]). In addition, it has demonstrated its theoretical flexibility: the structure is adaptable to admixture with alternative theories of ambiguity and unpredictability. Akram et al. [3–7], Chen et al. [24], Liu et al. [33], and Riaz et al. [50] have built hybrid structures that incorporate other notable properties of approximation knowledge structures. An N-soft structure [49] exists as a natural extension of soft topology and is a natural extension of topological studies [9, 54].

A. K. Das (✉)
Department of Mathematics, Bir Bikram Memorial College, Agartala 799004, India

C. Granados
Estudiante de Doctorado en Matemáticas, Magister en Ciencias Matemáticas, Universidad de Antioquia, Medellín, Colombia
e-mail: carlosgranadosortiz@outlook.es

S. Mukherjee et al. (eds.), *Intelligent Systems for Social Good*,
Advanced Technologies and Societal Change,
https://doi.org/10.1007/978-981-19-0770-8_11

The idea of the FST was started by Zadeh [59], thereafter, many new approaches and ideas have been offered to deal with imprecision and ambiguity, such as the hesitant fuzzy sets (HFSs) [55], MFSs [52], intuitionistic fuzzy set theory (IFST) [19], intuitionistic MFSs [25], IVFSs [62] and so on [1, 2, 20, 21, 43]. FST has a wide range of applications, including databases, neural systems, pattern recognition, medicine, fuzzy modelling, economics, and multicriteria DMPs (see [14, 18]). Maji et al. [40] described fuzzy soft set (FSS) in 2001. An FSS-based DMP was proposed by Roy and Maji [51]. Several authors later looked into the more broad properties and applications of FSS; for example, see [17, 22, 23, 26, 58]. Alcantud and Mathew (2017) [12] recently defined separable FSS and its use in DMPs with both positive and negative qualities. Alcantud et al. [10] developed a methodology for asset assessment using an FSS (flexible FSS) based decision-making technique. Atanassov [19] suggested the notion of IFST as a generalization of FST [59]. Maji et al. [41, 42] presented intuitionistic FSS (IFSS) as an important mathematical method for solving DMPs in an uncertain situation by combining SST [44] with IFST [19]. Several authors later looked into the more broad properties and applications of IFST and IFSS; for example, see [34, 53].

The topic of intertemporal FSS selection was first raised by Alcantud and Muoz Torrecillas [15]. The algorithms for IVFSSs in stochastic multi-criteria DMPs and neutrosophic soft DMM were introduced by Peng et al. [47, 48]. Furthermore, using WDBA and CODAS with the help of new data measures, Peng and Garg [46] suggested unique algorithms for IVFSSs in emergency DMPs. In the case of SST, Zhan and Alcantud [60] provided a rationalized assessment of the parameter reduction novel. One or more of the following constraints limited the majority of SST researchers (for example [37] or other updated hybrid model summaries): The evaluations can either be binary integers between 0 and 1, or real values between 0 and 1, such as FSSs or separable FSSs [40].

Both scenarios are discussed in [13], which includes an examination of partial data. In scenarios such as social assessment structures or the arrangement of ranking positions in ordinary life, however, we encounter data with a different framework which not binary. NSSs [31] are, nonetheless, the accurate formal expression of the idea of parameterized description of the universe of items based on a limited figure of ordered grades, and the other extended structures of SST that have been linked to social choice were mentioned by Fatimah et al. [29, 30]. The idea of parameter reduction in NSSs was recently presented by Akram et al. [8]. When the membership degrees of the alternatives are not uniquely defined, such as due to group knowledge or hesitancy [57, 61], HFSs [55] are useful. Hesitancy is a model that can be combined with other key structures (See [28, 35, 36, 45, 56] for contemporary examples).

Recently, Fatimah and Alcantud [27] introduced the concept of MFNSS and developed a DMM for solving DMPs, but in their DMM, there have some limitations. They used MFNSS evaluated by only one DM, so this method is may not be useful in the modelling of group-DMPs. The new structure combines the advantages of MFSs [52] with those of NSSs [31] and IVFSs [62], these two structures that have received a lot of attention in current years. The constructed method in this chapter is very advantageous for solving group-DMPs. To demonstrate the applicability of our methodology in practical situations, some examples are used.

2 Preliminary

Let us consider Ω represents the starting universe and Q represents a nonempty collection of parameters with $P \subseteq Q$. Let the power set of Ω is denoted as P(Ω). Let $N \in \{2, 3, 4, 5, ...\}$ and $R = \{0, 1, 2, 3, 4, 5, ..., N - 1\}$.

Definition 2.1 [59] A fuzzy set (FS) Z on Ω having a structure $Z = \{(o, \mu_Z(o)) : o \in \Omega\}$, where $\mu_z: \Omega \to [0, 1]$ is called membership function, and $\mu_z(o)$ is the membership value for each object $o \in \Omega$.

Assume that, in this chapter FS(Ω) means the collection of all FSs on Ω.

Definition 2.2 [44] A soft set over the nonempty universe Ω is a pair (ψ, P), where ψ is a mapping defined by ψ: P → P (Ω).

Definition 2.3 [53] An MFS Z on Ω is a set with a structure membership value for each object

where $\mu_k : \Omega \to [0, 1][0, 1]$ for $k = 1, 2, 3,\ldots,q$ are the real valued functions.

In this chapter, $MFs(\Omega)^q$ means the collection of all MFSs on Ω.

Definition 2.4 [31] A triple (Ψ, P, N) is called NSS on Ω, where $\Psi : P \to 2^{\Omega \times R}$ is a function, satisfying the condition, for each p ∈ P and o ∈ Ω there exists a unique couple $(o, r_p) \in \Omega \times R$ such that $(o, r_p) \in \Psi(p), r_p \in R$. The object o belongs to the collection of p-approximations of Ω with the grade r_p, according to the interpretation of the couple $(o, r_p) \in \Psi(p)$.

Definition 2.5 [5] A triple (ψ, P, N) is called a hesitant N-soft set (simply, H(N)SS) over Ω, where $\psi : P \to 2^{\Omega \times R}$ is a function such that for every p ∈ P, o ∈ Ω there exists at least one couple $(o, r_p) \in \Omega \times R$ such that $(o, r_p) \in \psi(p)$, $r_p \in R$.

Definition 2.6 [5] The collection h satisfying the condition $\phi \neq h \subseteq R = \{0, 1, 2, 3, 4, 5, ..., N - 1\}$ is said to be hesitant N-tuple (simply, H(N)T). Any H(N)SS has a tabular form consisting of a matrix with cells H(N)Ts.

Definition 2.7 [62] An IVFS Γ on Ω having the form $\Gamma = \{(o, [\mu_\Gamma^L(o), \mu_\Gamma^U(o)]) : u \in \Omega\}$, where $\mu_\Gamma^L, \mu_\Gamma^U : \Omega \to [0, 1]$ are functions.

IVFS(Ω) denote the collection of all IVFSs on Ω. Let Γ, Ψ ∈ IVFS(Ω). Then.
[i] Union of Γ and Ψ is denoted by Γ ∪ Ψ where

$$\Gamma \cup \psi = \{(o, [\max(\mu_\Gamma^L(o)\mu_\psi^L(o)), \max(\mu_\Gamma^U(o)\mu_\psi^U(o))]) : o \in \Omega\}$$

[ii] Intersection of Γ and Ψ is denoted as $\Gamma \cap \Psi$ where

$$\Gamma \cap \psi = \{(o, [\min(\mu_\Gamma^L(o)\mu_\psi^L(o)), \min(\mu_\Gamma^U(o)\mu_\psi^U(o))]) : o \in \Omega\}$$

3 d(N, q)-Set and Its Theoretical Analysis

In this chapter, we consider Ω represents the starting universe and Q represents a nonempty set of parameters. Let $P \subseteq Q$ and $X = \{p^{[\mu_X^L(p), \mu_X^U(p)]} : p \in P\}$, be an IVFS on P. Let $q, N \in \{2, 3, 4, 5, ...\}$ be two fixed numbers, where q is the dimension of our new structure and N distinguishes how many degrees of satisfaction with the parameters are permitted, allowing us to utilize $R = \{0, 1, 2, 3, 4, 5, ..., N-1\}$ as a collection of ordered grades.

Definition 3.1 Let us define $MFs(\Omega)^{(N,q)}$ as the set of all q-tuples of triples of objects from $R \times [0, 1] \times [0, 1]$ indexed by Ω, i.e., the collection of all objects having the structure $\{\langle o, (r_1(o), \mu_1(o)), (r_2(o), \mu_2(o)), ..., (r_q(o), \mu_q(o))\rangle : o \in \Omega\}$, where $r_k : \Omega \to R$, $\mu_k : \Omega \to [0, 1]$ are functions for $k = 1, 2, 3, \ldots, q$. A *d(N,q)*-set on Ω is a pair (Ψ, X) such that Ψ is a mapping $\Psi : X \to MFs(\Omega)^{(N,q)}$ defined by

$$\forall p^{[\mu_X^L(p), \mu_X^U(p)]} \in X,$$

$$\Psi\left(p^{[\mu_X^L(p), \mu_X^U(p)]}\right) = \{\langle o, (r_1(o), \mu_1(o)), (r_2(o), \mu_2(o)), ..., (r_q(o), \mu_q(o))\rangle : o \in \Omega\}.$$

Note: Simply, we denote the collection of all *d(N,q)*-sets on Ω by $d(\Omega, P)^{(N,q)}$ where the parameter set P is fixed.

Example 3.2 Let $\Omega = \{o_1, o_2, o_3\}$ be the universe of candidates and $P = \{pm_1, pm_2, pm_3\}$ is the set of attributes and $X = \{pm_1^{[0.5,0.6]}, pm_2^{[0.7,0.8]}, pm_3^{[0.6,0.7]}\}$ be an IVFS over P. A d(5,2)-set (Ψ, X) on Ω is defined by the assignments (Table 1).

$$\Psi(p_1^{[0.5,0.6]}) = \{\langle o_1, (3, 0.3), (4, 0.5)\rangle, \langle o_2, (2, 0.3), (4, 0.5)\rangle, \langle o_3, (2, 0.4), (3, 0.4)\rangle\}$$

Table 1 The *d(5,2)*-set (Ψ, *X*)

	pm_1 [0.5,0.6]	pm_2 [0.7,0.8]	pm_3 [0.6,0.7]
01	(3,0.3)(4,0.5)	(1,0.4)(3,0.5)	(3,0.4)(2,0.5)
02	(2,0.3)(4,0.5)	(2,0.3)(4,0.5)	(2,0.5)(4,0.6)
03	(2,0.4)(3,0.4)	(3,0.5)(2,0.5)	(2,0.5)(1,0.7)

$$\Psi(p_2^{[0.7,0.8]}) = \{\langle o_1, (1, 0.4), (3, 0.5)\rangle, \langle o_2, (2, 0.3), (4, 0.5)\rangle, \langle o_3, (3, 0.5), (2, 0.5)\rangle\}$$

$$\Psi(p_3^{[0.6,0.7]}) = \{\langle o_1, (3, 0.4), (2, 0.5)\rangle, \langle o_2, (2, 0.5), (4, 0.6)\rangle, \langle o_3, (2, 0.5), (1, 0.7)\rangle\}$$

Definition 3.3 Let us consider two d(N,q)-sets $(\psi, X), (\varphi, Y) \in d(\Omega, P)^{(N,q)}$, such that $\forall p^{[\mu_X^L(p),\mu_X^U(p)]} \in X, p^{[\mu_Y^L(p),\mu_Y^U(p)]} \in Y, \psi\left(p^{[\mu_X^L(p),\mu_X^U(p)]}\right) = \{\langle o, (r_1(o), \mu_1(o)), (r_2(o), \mu_2(o)), \ldots, (r_q(o), \mu_q(o))\rangle : o \in \Omega\}\varphi\left(p^{[\mu_Y^L(p),\mu_Y^U(p)]}\right) = \left\{\langle o, (r_1'(o), \mu_1'(o)), (r_2'(o), \mu_2'(o)), \ldots, (r_q'(o), \mu_q'(o))\rangle : o \in \Omega\right\}$.

Then we say that

[1] **Subset**: $(\psi, X)\tilde{\subseteq}(\varphi, Y)$ if

$$(i). \forall p \in P,\ \mu_X^L(p) \le \mu_Y^L(p) \text{ and } \mu_X^U(p) \le \mu_Y^U(p)]$$

$$(ii). \forall p \in P,\ \psi\left(p^{[\mu_X^L(p),\mu_X^U(p)]}\right) \subseteq \varphi\left(p^{[\mu_Y^L(p),\mu_Y^U(p)]}\right) \Leftrightarrow r_i(o) \le r_i'(o) \text{ and } \mu_i(o) \le \mu_i'(o)$$

$\forall o \in \Omega$, and $i = 1, 2, \ldots, q$

[2] **Union**: $(\psi, X)\tilde{\cup}(\varphi, Y) = (\rho, Z)$, where $\forall p^{[\mu_Z^L(p),\mu_Z^U(p)]} \in Z$,

$$\rho\left(p^{[\mu_Z^L(p),\mu_Z^U(p)]}\right) = \left\{\langle o, (r_1''(o), \mu_1''(o)), \ldots, (r_q''(o), \mu_q''(o))\rangle : o \in \Omega\right\}, \text{ where}$$

$\forall_o \in \Omega, r_i'(o) = \min\{r_i(o), r_i'(o)\}, \mu'' = \min\{\mu_i(o), \mu_i'(o)\}, i = 1, 2, \ldots, q$, and $\mu_Z^L(p) = \min\{\mu_X^L(p), \mu_Y^L(p)\}, \mu_Z^U(p) - \min\{\mu_X^U(p), \mu_Y^U(p)\}$

[3] **Intersection**: $(\psi, X)\tilde{\cap}(\varphi, Y) = (\rho, Z)$, where $\forall p^{[\mu_Z^L(p),\mu_Z^U(p)]} \in Z$,

$$\rho\left(p^{[\mu_Z^L(p),\mu_Z^U(p)]}\right) = \left\{\langle o, (r_1''(o), \mu_1''(o)), (r_2''(o), \mu_2''(o)), \ldots, (r_q''(o), \mu_q''(o))\rangle : o \in \Omega\right\},$$

$$\forall o \in \Omega,\ r_i''(o) = \min\{r_i(o), r_i'(o)\},\ \mu_i''(o) = \min\{\mu_i(o), \mu_i'(o)\},\ i = 1, 2, \ldots, q,$$

$$\text{and } \mu_Z^L(p) = \min\left\{\mu_X^L(p), \mu_Y^L(p)\right\},\ \mu_Z^U(p) = \min\left\{\mu_X^U(p), \mu_Y^U(p)\right\}.$$

Definition 3.4 We consider a *d(N,q)*-set $(\Psi, X) \in d(\Omega, P)^{(N,q)}$. Its induced interval-valued fuzzy parameterized hesitant N-soft set of dimension q (briefly, *h(N,q)*-set) is the pair (H_Ψ, X), where $H_\Psi : X \to P(R)$ is a mapping, such that $H_\Psi(p_j^{[\mu_X^L(p),\mu_X^U(p)]}) = \left\{\left\langle o_i, \left\{r_1^j(o_i), r_2^j(o_i), \ldots, r_q^j(o_i)\right\}\right\rangle : o_i \in \Omega\right\}$, $\forall p^{[\mu_X^L(p),\mu_X^U(p)]} \in X$.

Definition 3.5 Let us fix $(\Psi, X) \in d(\Omega, P)^{(N,q)}$ and $\alpha = (\alpha_1, \alpha_2, \ldots, \alpha_m)$, a vector of thresholds with $\alpha_i \in [0, 1]$, i = 1, 2,…m. Then the *h(N,q,α)*-set induced from (Ψ, X) is the triple (H_Ψ, X, α), where $H_\Psi : X \rightarrow P(R)$ is a mapping, such that $H_\Psi(p_j^{[\mu_X^L(p),\mu_X^U(p)]}) = \left\{\left\langle o_i, \left\{r_t^j(o_i) : \mu_t^j(o_i) \geq \alpha_j, t = 1, 2, ..., q\right\}\right\rangle : o_i \in \Omega\right\}$.

Definition 3.6 Let X be an IVFS and $\delta, \gamma \in [0, 1]$ such that $\delta+\gamma= 1$. Then a soft fuzzification operator $S_{(\delta,\gamma)}$ on IVFS X is an FS denoted as $S_{(\delta,\gamma)}(X)$ and defined by $S_{(\delta,\gamma)}(X) = \left\{p^{\mu_{S_{(\delta,\gamma)}(X)}(p)} : p \in P\right\}$, where $\mu_{S_{(\delta,\gamma)}(X)}(p) = \delta\mu_X^L(p) + \gamma\mu_X^U(p)$.

By adjusting the value of γ, δ an IVFS can be transformed into any FS. Specially, if $\gamma = 0$, $\delta = 1$, then the pessimistic-pessimistic FS, defined by $S_{(1,0)}(X) = \left\{p^{\mu_X^L(p)} : p \in P\right\}$.

If $\gamma = 1$, $\delta = 0$, then the optimistic-optimistic FS, defined by $S_{(0,1)}(X) = \left\{p^{\mu_X^U(p)} : p \in P\right\}$.

If $\gamma = 0.5$, $\delta = 0.5$, then the neutral–neutral FS, defined by $S_{(0.5,0.5)}(X) = \left\{p^{\frac{\mu_X^L(p)+\mu_X^U(p)}{2}} : p \in P\right\}$.

4 DMM Based on $d(N,q)$-Sets and $h(N,q,\alpha)$-Sets

Now, we present our machine learning algorithm for solving group DMPs based on d(N,q)-sets and *h(N,q,α)*-sets. The steps of our proposed DMM are listed below:

Algorithm 1

Step 1: Enter a nonempty universe $\Omega = \{o_1, o_2, ..., o_n\}$, a set of parameters $P = \{p_1, p_2, ..., p_m\}$ and a group of DMs {M1, M2,…, Mk}.
Step 2: Enter the corresponding decision maker's opinions (*d(N,q)*-sets) $(\Psi_1, X_1), (\Psi_2, X_2), ..., (\Psi_k, X_k)$ respectively.
Step 3: Enter vector α^λ of thresholds corresponding to each DM M_λ, $\lambda = 1,2,\ldots,k$, i.e., a threshold $\alpha_j^\lambda \in [0, 1]$ with respect to every parameter $p_j \in P$.
Step 4: Obtain the resultant *d(N,q)*-set (Ψ, X) (using union or intersection) and resultant threshold $\alpha = \left\{\alpha_j \in [0, 1],\ j = 1, 2, 3, ..., m\right\}$ where α_j associated with each attribute $p_j \in P$(using any FS operation).
Step 5: Obtain the *h(N,q,α)*-set $\left(H_\Psi, X, \alpha\right)$ in its tabular form.
Step 6: Obtain the scores $\Gamma\left(h_j(o_i)\right)$ of all the H(N)Ts by taking any operation (say, arithmetic mean, geometric mean, or harmonic mean), $\forall o_i \in \Omega$, and j = 1, 2, ….., m.
Step 7: Enter the values of and $\delta, \gamma \in [0, 1]$ such that $\delta+\gamma= 1$ obtain $S_{(\delta,\gamma)}(X)$.
Step 8: Compute $u_i = \sum_{j=1}^{m} \mu_{S_{(\delta,\gamma)}(X)}(p_j)\times\Gamma\left(h_j(o_i)\right)$, $\forall o_i \in \Omega$.
Step 9: The best optimal choice is to select o_s if u_s is maximized.
Step 10: If o_s has many values, any of o_s may be selected.

Remark 4.1 In the 9th-step of our constructed DMM, one can return to the 3rd or 4th steps or 7th step and change the thresholds α^{λ} or operations or the values of γ, δ respectively that he previously used to change the optimal choice, particularly when there are lots of optimal decisions to choose from.

5 Result and Discussions

Example 5.1 Let $\Omega = \{o_1, o_2, o_3, o_4\}$ be the universe of candidates, and $P = \{pm_1, pm_2, pm_3\}$ is the set of attributes. We suppose that there are two DMs such as D1, D2 and their observations $(\psi, X), (\varphi, Y) \in d(\Omega, P)^{(5,3)}$ as in Tables 2 and 3 respectively. Also, let {0.5, 0.5, 0.6} and {0.4, 0.4, 0.5} be thresholds corresponding to the DMs D1 and D2 respectively. The resultant *d(5,3)*-set (Ψ, *Z*) (using union) as in Table 4 and the resultant threshold is α = {0.5, 0.5, 0.6} (using FS-union). We obtain h(5, 3, α) as shown in Table 5. We use the arithmetic score on H(N)Ts. Table

Table 2 The *d*(5,3)-set (ψ, X)

Ω	pm_1 [0.4,0.5]	pm_2 [0.5,0.6]	pm_3 [0.7,0.8]
01	(3,0.3)(4,0.5)(2,0.6)	(1,0.4)(4,0.3)(2,0.5)	(2,0.5)(3,0.3)(2,0.6)
02	(2,0.3)(1,0.5)(3,0.5)	(2,0.3)(4,0.5)(3,0.5)	(3,0.6)(4,0.4)(2,0.4)
03	(2,0.4)(3,0.4)(1,0.5)	(1,0.5)(2,0.5)(4,0.4)	(3,0.5)(2,0.3)(1,0.5)
04	(1,0.5)(3,0.6)(3,0.4)	(1,0.4)(2,0.6)(3,0.4)	(4,0.3)(2,0.6)(3,0.4)

Table 3 The *d(5,3)*-set (φ, Y)

Ω	pm_1 [0.5,0.6]	pm_2 [0.6,0.7]	pm_3 [0.6,0.7]
01	(2,0.4)(3,0.6)(2,0.5)	(3,0.4)(2,0.4)(1,0.6)	(4,0.4)(2,0.5)(1,0.5)
02	(3,0.4)(4,0.5)(3,0.6)	(3,0.4)(3,0.6)(2,0.4)	(3,0.5)(4,0.5)(2,0.6)
03	(2,0.3)(4,0.4)(1,0.4)	(3,0.5)(4,0.4)(2,0.4)	(4,0.4)(2,0.4)(3,0.6)
04	(2,0.4)(4,0.5)(3,0.5)	(4,0.6)(2,0.5)(3,0.6)	(4,0.5)(2,0.5)(3,0.4)

Table 4 The *d(5,3)*-set (Ψ, *X*)

Ω	pm_1 [0.5,0.6]	pm_2 [0.6,0.7]	pm_3 [0.7,0.8]
01	(3,0.4)(4,0.6)(2,0.6)	(3,0.4)(4,0.4)(2,0.6)	(4,0.5)(3,0.5)(2,0.6)
02	(3,0.4)(4,0.5)(3,0.6)	(3,0.4)(4,0.6)(3,0.5)	(3,0.6)(4,0.5)(2,0.6)
03	(2,0.4)(4,0.4)(1,0.5)	(3,0.4)(4,0.5)(4,0.4)	(4,0.5)(2,0.4)(3,0.6)
04	(2,0.5)(4,0.6)(3,0.5)	(4,0.6)(2,0.6)(3,0.6)	(4,0.5)(2,0.6)(3,0.4)

Table 5 The *h(5,3,α)*-set (H_Ψ, X, α)

Ω	pm_1 [0.5,0.6]	pm_2 [0.6,0.7]	pm_3 [0.7,0.8]
01	{4, 2}	{2}	{2}
02	{4, 3}	{4, 3}	{2}
03	{1}	{3, 4}	{3}
04	{2, 4, 3}	{4, 2, 3}	{2}

Table 6 $S_{(1,0)}(X)$ and the scores $\Gamma(h_j(o_i))$ with u_i

Ω	pm_1 0.5	pm_2 0.6	pm_3 0.7	u_i
01	3	2	2	4.1
02	3.5	3.5	2	5.25
03	1	3.5	3	4.7
04	4.5	4.5	2	6.35

6 shows the results of the computations at steps 6, 7, and 8. Step 9 suggests that the option 04 be chosen in the last column.

Another example is one in which the constituent pieces have distinct specifications. This demonstrates Algorithm 1 adaptability and versatility.

6 Comparison Analyses

The DMM given in [27] is good for solving DMPs, but in their DMM they used MFNSS evaluated by only one DM, so these methods are may not be useful in the modelling of group-DMPs, but the constructed method in this chapter is very advantageous for solving group-DMPs. Also, the importance of the weights of parameters is considered by only one DM in [27], but in our constructed DMM, every DM can consider their own weights with the parameters so that every DM can give the importance of parameter selections according to their choice.

7 Conclusions

IVFSs can effectively represent and simulate the uncertainty and diversity of judgment information offered by DMs. In comparison to FSs, IVFSs are highly beneficial for expressing vagueness and uncertainty more accurately. As a result, in this research work, we offer an approach for solving group DMPs with d(N,q)-sets by extending the MFNSS based DMM. The DMM given in [27] is good for solving DMPs, but in their DMM they used MFNSS evaluated by only one DM, so these methods are may not be useful in the modelling of group-DMPs, but the constructed method in this chapter is very advantageous for solving group-DMPs. Also, the importance of the weights of parameters is considered by only one DM in [27], but in our constructed DMM, every DM can consider their own weights with the parameters so that every DM can give the importance of parameter selections according to their choice.

References

1. Abdulkareem, K.H., Arbaiy, N., Zaidan, A.A., et al.: A new standardisation and selection framework for real-time image dehazing algorithms from multi-foggy scenes based on fuzzy Delphi and hybrid multi-criteria decision analysis methods. Neural Comput. Appl. https://doi.org/10.1007/s00521-020-05020-4 (2020)
2. Abdulkareem, K.H., Arbaiy, N., Zaidan, A.A, et al.: A novel multiperspective benchmarking framework for selecting image dehazing intelligent algorithms based on BWM and group VIKOR techniques. Int. J. Inf. Technol. Decis. Making **19**(3), 909–957 (2020)
3. Akram, M., Adeel, A.: TOPSIS approach for MAGDM based on interval-valued hesitant fuzzy N-soft environment. Int. J. Fuzzy Syst. **21**(3), 993–1009 (2019)
4. Akram, M., Adeel, A., Alcantud, J.C.R.: Group decision-making methods based on hesitant N-soft sets. Expert Syst. Appl. **115**, 95–105 (2019)
5. Akram, M., Adeel, A., Alcantud, J.C.R.: Hesitant fuzzy N-soft sets: a new model with applications in decision-making. J. Intell. Fuzzy. Syst. **36**(6), 6113–6127 (2019)
6. Akram, M., Adeel, A., Alcantud, J.C.R.: Fuzzy N-soft sets: a novel model with applications. J. Intell. Fuzzy Syst. **35**(4), 4757–4771 (2018)
7. Akram, M., Ali, G., Alcantud, J.C.R.: New decision-making hybrid model: intuitionistic fuzzy N-soft rough sets. Soft Comput. **23**(20), 9853–9868 (2019)
8. Akram, M., Ali, G., Alcantud, J.C.R., Fatimah, F.: Parameter reductions in N-soft sets and their applications in decision-making. Expert Syst (2020)
9. Alcantud, J.C.R.: Soft open bases and a novel construction of soft topologies from bases for topologies. Mathematics **8**(5), 672 (2020)
10. Alcantud, J.C.R., Cruz-Rambaud, S., Torrecillas, M.J., Muñoz,: Valuation fuzzy soft sets: a flexible fuzzy soft set based decision making procedure for the valuation of assets. Symmetry **9**, 253 (2017)
11. Alcantud, J.C.R., Feng, F., Yager, R.R.: An N-soft set approach to rough sets. IEEE Trans. Fuzzy Syst. **28**(11), 2996–3007 (2020)
12. Alcantud, J.C.R., Mathew, T.J.: Separable fuzzy soft sets and decision making with positive and negative attributes. Appl. Soft Comput. **59**, 586–595 (2017)
13. Alcantud, J.C.R., Santos-García, G.: A new criterion for soft set based decision-making problems under incomplete information. Int. J. Comput. Intell. Syst. **10**, 394–404 (2017)
14. Alcantud, J.C.R., Torra, V.: Decomposition theorems and extension principles for hesitant fuzzy sets. Inf. Fus. **41**, 48–56 (2018)
15. Alcantud, J.C.R., Muñoz Torrecillas, M.J.: Intertemporal choice of fuzzy soft sets. Symmetry **9**, 253 (2017)
16. Ali, M.I., Feng, F., Liu, X.Y., Min, W.K., Shabir, M.: On some new operations in soft set theory. Comput. Math. Appl. **57**(9), 1547–1553 (2009)
17. Al-Qudah, Y., Hassan, N.: Complex multi-fuzzy soft set: its entropy and similarity measure. IEEE Access **6**, 65002–65017 (2018)
18. Al-Qudah, Y., Hassan, N.: Operations on complex multi-fuzzy sets. J. Intell. Fuzzy Syst. **33**, 1527–1540 (2017)

19. Atanassov, K.T.: Intuitionistic fuzzy sets. Fuzzy Sets Syst. **20**(1), 87–96 (1986)
20. Azam, M., Bouguila, N.: Bounded generalized Gaussian mixture model with ICA. Neural Process Lett. **49**, 1299–1320 (2019)
21. Azam, M., Bouguila, N.: Multivariate bounded support Laplace mixture model. Soft Comput. **24**, 13239–13268 (2020)
22. Çağman, N., Çitak, F., Enginoğlu, S.: Fuzzy parameterized fuzzy soft set theory and its applications. Turk. J. Fuzzy Syst. **1**(1), 21–35 (2010)
23. Çağman, N., Çitak, F., Enginoğlu, S.: FP-soft set theory and its applications. Ann. Fuzzy Math. Inf. **2**(2), 219–226 (2011)
24. Chen, Y., Liu, J., Chen, Z., Zhang, Y.: Group decision-making method based on generalized vague N-soft sets. In: Chinese Control And Decision Conference (CCDC), pp. 4010–4015 (2020)
25. Das, S., Kar, S.: Intuitionistic multi fuzzy soft set and its application in decision making. In: Maji P., Ghosh A., Murty M.N., Ghosh K., Pal S.K. (eds) Pattern Recognition and Machine Intelligence. PReMI 2013. Lecture Notes in Computer Science, vol. 8251. Springer, Berlin, Heidelberg. https://doi.org/10.1007/978-3-642-45062-4_82(2013)
26. Dey, A., Pal, M.: Generalised multi-fuzzy soft set and its application in decision making. Pacif. Sci. Rev. A Nat. Sci. Eng. **17**(1), 23–28 (2015)
27. Fatimah, F., Alcantud, J.C.R.: The multi-fuzzy N-soft set and its applications to decision-making. Neural Comput. Appl. (2021). https://doi.org/10.1007/s00521-020-05647-3
28. Fatimah, F., Alcantud, J.C.R.: Expanded dual hesitant fuzzy sets. In: International Conference on Intelligent Systems (IS), , pp 102–108. https://doi.org/10.1109/IS.2018.8710539(2018)
29. Fatimah, F., Rosadi, D., Hakim, R.B.F., Alcantud, J.C.R.: (2019) Probabilistic soft sets and dual probabilistic soft sets in decision-making. Neural Comput. Appl. **31**(Suppl 1:397), 397–407
30. Fatimah, F., Rosadi, D., Hakim, R.B.F.: Probabilistic soft sets and dual probabilistic soft sets in decision making with positive and negative parameters. J. Phys. Conf. Ser. **983**(1), 012112 (2018)
31. Fatimah, F., Rosadi, D., Hakim, R.B.F., Alcantud, J.C.R.: N-soft sets and their decision-making algorithms. Soft Comput. **22**(12), 3829–3842 (2018)
32. Kamacı, H., Petchimuthu, S.: Bipolar N-soft set theory with applications. Soft Comput. **24**, 16727–16743 (2020)
33. Liu, J., Chen, Y., Chen, Z., Zhang, Y.: Multi-attribute decision making method based on neutrosophic vague N-soft sets. Symmetry **12**, 853 (2020)
34. Liu, X., Kim, H., Feng, F., Alcantud, J.C.R.: Centroid transformations of intuitionistic fuzzy values based on aggregation operators. Mathematics **6**(11), 215 (2018)
35. Liu, P., Zhang, L.: An extended multiple criteria decisionmaking method based on neutrosophic hesitant fuzzy information. J. Intell. Fuzzy Syst. **32**(6), 4403–4413 (2017)
36. Liu, P., Zhang, L.: Multiple criteria decision-making method based on neutrosophic hesitant fuzzy Heronian mean aggregation operators. J. Intell. Fuzzy Syst. **32**(1), 303–319 (2017)
37. Ma, X., Liu, Q., Zhang, J.: A survey of decision-making methods based on certain hybrid soft set models. Artif. Intell. Rev. **47**(4), 507–530 (2017)
38. Maji, P.K., Biswas, R., Roy, A.R.: Soft set theory. Comput. Math. Appl. **45**(4–5), 555–562 (2003)
39. Maji, P.K., Biswas, R., Roy, A.R.: An application of soft sets in decision-making problem. Comput. Math. Appl. **44**(8–9), 1077–1083 (2002)
40. Maji, P.K., Biswas, R., Roy, A.R.: Fuzzy soft sets. J. Fuzzy Math. **9**(3), 589–602 (2001)
41. Maji, P.K., Biswas, R., Roy, A.R.: Intuitionistic fuzzy soft sets. J. Fuzzy Math. **9**(3), 677–692 (2001)
42. Maji, P.K., Roy, A.R., Biswas, R.: On intuitionistic fuzzy soft sets. J. Fuzzy Math. **12**(3), 669–683 (2004)
43. Mohammed, M.A., Abdulkareem, K.H., et al.: Benchmarking methodology for selection of optimal COVID-19 diagnostic model based on entropy and TOPSIS methods. IEEE Access. **8**, 99115–99131 (2020)
44. Molodtsov, D.: Soft set theory-first results. Comput. Math. Appl. **37**(4–5), 19–31 (1999)

45. Peng, X., Dai, J.: Hesitant fuzzy soft decision-making methods based on WASPAS, MABAC and COPRAS with combined weights. J. Intell. Fuzzy Syst. **33**(2), 1313–1325 (2017)
46. Peng, X.D., Garg, H.: Algorithms for interval-valued fuzzy soft sets in emergency decision-making based on WDBA and CODAS with new information measure. Comput. Ind. Eng. **119**, 439–452 (2018)
47. Peng, X.D., Liu, C.: Algorithms for neutrosophic soft decision making based on EDAS, new similarity measure and level soft set. J. Intell. Fuzzy Syst. **32**(1), 955–968 (2017)
48. Peng, X.D., Yang, Y.: Algorithms for interval-valued fuzzy soft sets in stochastic multi-criteria decision-making based on regret theory and prospect theory with combined weight. Appl. Soft Comput. **54**, 415–430 (2017)
49. Riaz, M., Çagman, N., Zareef, I., Aslaam, M.: N-soft topology and its applications to multi-criteria group decision making., J. Intell. Fuzzy Syst. **36**(6), 6521–6536 (2019)
50. Riaz, M., Naeem, K., Zareef, I., Afzal, D.: Neutrosophic N-soft sets with TOPSIS method for multiple attribute decision making. Neutrosophic Sets an Syst. **32**, 1–23 (2020)
51. Roy, A.R., Maji, P.K.: A fuzzy soft set theoretic approach to decision making problems J. . Comput. Appl. Math. **203**, 412–418 (2007)
52. Sebastian, S., Ramakrishnan, T.V.: Multi-fuzzy sets: an extension of fuzzy sets. Fuzzy Inf. Eng. **1**, 35–43 (2011)
53. Shinoj, T.K., John, S.J.: Intuitionistic fuzzy multisets and its application in medical diagnosis. World Acad. Sci. Eng. Technol. **61**, 1178–1181 (2012)
54. Terepeta, M.: On separating axioms and similarity of soft topological spaces. Soft Comput. **23**(3), 1049–1057 (2019)
55. Torra, V.: Hesitant fuzzy sets. Int. J. Intell. Syst. **25**(6), 529–539 (2010)
56. Torra, V., Narukawa, Y.: On hesitant fuzzy sets and decisions. IEEE Int. Conf. Fuzzy Syst. **1–3**, 1378–1382 (2009)
57. Xia, M.M., Xu, Z.S.: Hesitant fuzzy information aggregation in decision-making. Int. J. Appr. Reason. **52**, 395–407 (2011)
58. Yang, Y., Tan, X., Meng, C.: The multi-fuzzy soft set and its application in decision making. Appl. Math. Model. **37**, 4915–4923 (2013)
59. Zadeh, L.A.: Fuzzy sets., Inf. Control. **8**(3), 338–353 (1965)
60. Zhan, J., Alcantud, J.C.R.: A survey of parameter reduction of soft sets and corresponding algorithms. Artif. Intell. Rev. **52**(3), 1839–1872 (2019)
61. Zhu, B., Xu, Z.S., Xu, J.P.: Deriving a ranking from hesitant fuzzy preference relations under group decision-making. IEEE Trans. Cybern. **44**(8), 1328–119 (2014)
62. Gorzałczany, M.B.: A method of inference in approximate reasoning based on interval valued fuzzy sets. Fuzzy Sets Syst. **21**, 1–17 (1987)

An Innovative Approach to Fuzzy Soft Set Based Investment Framework Using Machine Learning Algorithm

Ajoy Kanti Das and Carlos Granados

1 Introduction

Useful applications of soft set theory (SST) [20], mainly its use in economics, computational intelligence, and data engineering, have gained prominence. Since classical mathematical techniques are insufficient to explain many complex problems in these fields, vague principles have been used in recent years in various fields such as information technology, computer applications, pharmacology, medicine, engineering, etc. Since the classification of objects in SST [20] is not constrained, researchers may select the various types of parameters they want, significantly simplifying the decision-making (DM) process and making it more capable in the lack of incomplete knowledge. While several mathematical methods for dealing with uncertainties are available, such as fuzzy set [34], probability theory, interval-valued mathematics, and so on, each of these techniques has its own set of challenges. Furthermore, all of these methods are deficient in the parameterization of tools, which means they cannot be used to solve problems, especially in the economic, environmental, and social realms. In the sense that it is clear of the aforementioned difficulties, SST [20] stands out. Molodstov [20] proposed soft set theory (SST) as a numerical method for modeling with uncertainties. Maji et al. [16] went on to present several new concepts on SST, such as intersection, union, complements, and subset, as well as a detailed discussion of the use of SST in DMPs. Ali et al. [1] provided several definitions on SST and shown that De Morgan's conditions satisfied in SST to these novel operations. Maji et al. [17] had shown the applications based on SST in DMPs. Thereafter,

A. K. Das (✉)
Department of Mathematics, Bir Bikram Memorial College, Agartala 799004, India

C. Granados
Estudiante de Doctorado en Matemáticas, Magister en Ciencias Matemáticas, Universidad de Antioquia, Medellín, Colombia
e-mail: carlosgranadosortiz@outlook.es

S. Mukherjee et al. (eds.), *Intelligent Systems for Social Good*,
Advanced Technologies and Societal Change,
https://doi.org/10.1007/978-981-19-0770-8_12

several researchers doing their innovative research work in this theory and applied in various field. Recently, Zhan and Alcantud [35] introduced the concept of soft rough covering and demonstrated how it can be used in DMPs. Rajput et al. [28] defined soft almost ββ-continuity in soft topological spaces. Dalkılıç [4] introduced a novel approach to SST-based DM under uncertainty.

Since Zadeh [34] introduced the idea of fuzzy sets, several new approaches and theories for dealing with imprecision and ambiguity have been proposed. Deng and Jiang [8] presented the D number-based game theory and adversarial DM in a fuzzy situation. Mandal and Ranadive [18] introduced the concept of multi granulation fuzzy probabilistic rough sets and their applications in DM. Thereafter, Wan Mohd et al. [33] presented a DM method using the Pythagorean fuzzy sets. Li et al. [14] described the method of uncertainty using the Gaussian kernel. Maji et al. [15] described FSS by combining soft set [20] and fuzzy set [34], which have a lot of prospective for solving DMPs. The applications of FSS theory have been gradually concentrated by using these concepts. Roy and Maji [29] first proposed FSS base DM. Thereafter, Kong et al. [12] provided an FSS theoretic approach to DMPs and Feng et al. [9] introducing an advanced and adjustable method to solve FSS-based DMPs. Thereafter, Zhu and Zhan [36] described and presented the t-norm operations on fuzzy parameterized FSSs, as well as their applications in DM. Vimala et al. [32] studied the fuzzy soft cardinality in lattice ordered fuzzy soft groups as well as shown their applications in DMPs. Peng and Li [26] proposed an algorithm for hesitant fuzzy soft DM using revised aggregation operations. Shakila and Selvi [30] had written a note on fuzzy soft paraopen sets and maps in fuzzy soft topological space (FSTS) and Nihal [23] studied the concept of pasting lemma on an FSTS with mixed structure. Lathamaheswari et al. [13] provided the concept of triangular interval type-2 FSS and also shown its applications. Petchimuthu et al. [27] defined the generalized products and mean operations of fuzzy soft matrices as well as discussed their applications in DMPs. Smarandache et al. [31] reviewed the concepts FSTS, intuitionistic FSTS, and neutrosophic soft topological spaces. Paik and Mondal [24] introduced a distance-similarity technique using fuzzy sets and FSSs. Chinram et al. [3] introduced some geometric aggregation functions under q-rung orthopair FSS as well as their applications in MCDM. Deli and Çağman [7] presented intuitionistic fuzzy parameterized soft set theory and its application in DM and Jamkhaneh and Nadarajah [11] developed a new generalized intuitionistic fuzzy set. Mukherjee and Das [21, 22] proposed fuzzy soft multi sets based DM problems as well as interval valued intuitionistic fuzzy soft set in investment DM. Paik and Mondal [25] had shown the applications of FSSs in a type-2 environment. Močkoř and Hurtik [19] used the concept FSSs in image processing applications. Gao and Wu [10] defined filter and also applied in FSTSs. Dalkılıç and Demirtaş [5] introduced the idea of bipolar fuzzy soft D-metric spaces. Dalkiliç [6] defined topology on virtual fuzzy parametrized-FSSs. Bhardwaj and Sharma [2] described an advanced uncertainty measure using FSSs and shown its application in DMPs.

In this research, we have introduced a new approach focused on FSS for an investing plan in a risky situation. First, we will go through some concepts and outcomes that will help us continue our discussion, then we have defined TNP and

TCP on FAVSs of an FSS, and applying these two products we present a strategy based on an FSS. In the end, we discuss the use of an FSS in IDM and numerical examples demonstrate the viability of our proposed FSS-based strategy in practice.

2 Preliminary

In this part, we will go through some basic concepts and outcomes that will help us think more deeply. Let V represents a universe and Q represents a collection of decision attributes with D ⊆ Q. Also, let P(V) means the power set of V.

Definition 2.1 [34] A fuzzy set (simply, FS) Y on V is of the form $Y = \{(d,\ \mu_Y(d)) : d \in V\}$, where $\mu_Y : V \to [0,\ 1]$ is called the membership function and $\mu_Y(d)$ means the membership value (MV) of $d \in V$.

If $\forall d \in V$, $\mu_Y(d) = 1$, then Y becomes a crisp set. Let FS(V) means the collection of all fuzzy subsets of V and $X_1, X_2,\ X_3, X_4 \in FS(V)$.

Definition 2.2 [36] t-norm is monotonic, associative, and commutative function t: [0,1] × [0,1] → [0,1] such that

(a) $t(0, 0) = 0$ and $t(1, \mu_{X_1}(d)) = \mu_{X_1}(d),\ d \in V$;

(b) If $\mu_{X_1}(d) \le \mu_{X_3}(d)$ and $\mu_{X_2}(d) \le \mu_{X_4}(d)$, then $t(\mu_{X_1}(d), \mu_{X_2}(d)) \le t(\mu_{X_3}(d), \mu_{X_4}(d))$;

(c) $t(\mu_{X_1}(d), \mu_{X_2}(d)) = t(\mu_{X_2}(d), \mu_{X_1}(d))$;

(d) $t(\mu_{X_1}(d), t(\mu_{X_2}(d), \mu_{X_3}(d))) = t(t(\mu_{X_1}(d), \mu_{X_2}(d)), \mu_{X_3}(d))$.

Definition 2.3 [36] t-conorm is associative, monotonic, and commutative function s: [0,1] × [0,1] → [0,1], such that

(a) $s(1, 1) = 1$ and $s(0, \mu_{X_1}(d)) = s(\mu_{X_1}(d), 0) = \mu_{X_1}(d),\ d \in V$;

(b) If $\mu_{X_1}(d) \le \mu_{X_3}(d)$ and $\mu_{X_2}(d) \le \mu_{X_4}(d)$, then $s(\mu_{X_1}(d), \mu_{X_2}(d)) \le s(\mu_{X_3}(d), \mu_{X_4}(d))$;

(c) $s(\mu_{X_1}(d), \mu_{X_2}(d)) = s(\mu_{X_2}(d), \mu_{X_1}(d))$;

(d) $s(\mu_{X_1}(d), s(\mu_{X_2}(d), \mu_{X_3}(d))) = s(s(\mu_{X_1}(d), \mu_{X_2}(d)), \mu_{X_3}(d))$.

Definition 2.4 [20] Soft set on V refers to a couple (*F, D*), where *F: D* → *P(V)* is a function.

Definition 2.5 [15] A pair (ψ, D) is said to be an FSS over V, where $\psi: D \to FS(V)$ is a mapping.

The FAVS of the parameter o is referred to as ψ(o) for any o ϵ D, and it can be written as

$$\Psi(o) = \left\{ \frac{d}{\mu_{\Psi(o)}(d)} : d \in V \right\},$$

where $\mu_{\Psi(o)}(d)$ is the fuzzy MV of the element (object) d holds on the parameter $o \in D$.

Simply, $FAVS(\Psi, D) = \{\Psi(o) : o \in D\}$.

3 TNP and TCP on $FAVS(\Psi, D)$

In this part, we define the *TNP* and *TCP* on $FAVS(\Psi, D)$. Let (Ψ, D) be an FSS over (V, Q) and let $\Psi(o_1), \Psi(o_2), ..., \Psi(o_n) \in FAVS(\Psi, D)$.

Definition 3.1 *TNP* is denoted by $\Psi(o_1) \otimes \Psi(o_2) \otimes .... \otimes \Psi(o_n)$ and defined as

$$\Psi(o_1) \otimes \Psi(o_2) \otimes .... \otimes \Psi(o_n) = \left\{ \frac{d}{\mu_{\Psi(o_1)\otimes\Psi(o_2)\otimes....\otimes\Psi(o_n)}(d)} : d \in V \right\}$$

where

$$\mu_{\Psi(o_1)\otimes\Psi(o_2)\otimes....\otimes\Psi(o_n)}(d) = \frac{\mu_{\Psi(o_1)}(d).\mu_{\Psi(o_2)}(d)...\mu_{\Psi(o_n)}(d)}{n - [\mu_{\Psi(o_1)}(d) + \mu_{\Psi(o_2)}(d) + ... + \mu_{\Psi(o_n)}(d) - \mu_{\Psi(o_1)}(d).\mu_{\Psi(o_2)}(d)...\mu_{\Psi(o_n)}(d)]}$$

Definition 3.2 *TCP* is denoted by $\Psi(o_1) \oplus \Psi(o_2) \oplus .... \oplus \Psi(o_n)$ and defined as

$$\Psi(o_1) \oplus \Psi(o_2) \oplus .... \oplus \Psi(o_n) = \left\{ \frac{d}{\mu_{\Psi(o_1)\oplus\Psi(o_2)\oplus....\oplus\Psi(o_n)}(d)} : d \in V \right\}$$

where

$$\mu_{\Psi(o_1)\oplus\Psi(o_2)\oplus....\oplus\Psi(o_n)}(d) = \frac{\mu_{\Psi(o_1)}(d) + \mu_{\Psi(o_2)}(d) + ... + \mu_{\Psi(o_n)}(d)}{n + \mu_{\Psi(o_1)}(d).\mu_{\Psi(o_2)}(d)...\mu_{\Psi(o_n)}(d)}$$

4 FSS-Based IDM

In this present sec., we propose our IDM based on FSS.

Algorithm 1

Step 1. Enter the (resultant) FSS (Ψ, D).

Step 2. Enter the preference of investment-factors $o_1, o_2, ..., o_n \in D$ by the investor.

Step 3. Obtain the *TNP* (or *TCP*) on $\Psi(o_1), \Psi(o_2), ..., \Psi(o_n) \in FAVS(\Psi, D)$.

Step 4. The optimal decision is to select d_k if the membership value (MV) of d_k i.e. $\mu_{\Psi(o_1)\otimes\Psi(o_2)\otimes....\otimes\Psi(o_n)}(d_k)$ is maximized.

Step 5. If d_k has more than one value then any one of d_k may be chosen.

5 Result and Discussions

To apply the concept of *FSS* in IDM problems, we select the following investment-factors which manipulate the investment plan and different avenues that prefer by an investor.

5.1 Investment-Factors (IFs)

Following are the most important investment-factors that influence the investment plan:

IF1 = Interest rates
IF2 = High returns
IF3 = Stable return
IF4 = Minimum risk
IF5 = Tax concession
IF6 = Easy accessibility
IF7 = Max profit in short time
IF8 = Safety of funds
IF9 = Technological developments
IF10 = Government policy

5.2 Investment-Avenues (IAs):

Some most important avenues that are mostly preferred by investors are as follows:

IA_1—Recurring Deposits
IA_2—Public Provident Fund (PPF)
IA_3—National Pension Scheme (NPS)
IA_4—Gold
IA_5—Mutual Fund
IA_6—Shares and Stocks
IA_7—Bank Deposit
IA_8—Postal Savings
IA_9—Insurance
IA_{10}—Employee Provident Fund (EPF)

To apply FSS in IDM, we have considered the different avenues as a universal set $V = \{IA_1, IA_2, IA_3, IA_4, IA_5, IA_6\}$ and the factors as a set of parameters Q = {IF1, IF2, IF3, IF4, IF5, IF6, IF7, IF8, IF9, IF10} and let D = {IF1, IF2, IF3, IF4, IF5} $\subseteq$ Q.

Also, we consider an FSS (Ψ, *D*) as in Table 1.

Table 1 The FSS (Ψ, D)

V	ψ(IF1)	ψ(IF2)	ψ(IF3)	ψ(IF4)	ψ(IF5)
IA_1	0.85	0.5	0.45	0.85	0.75
IA_2	0.8	0.6	0.35	0.8	0.9
IA_3	0.85	0.5	0.45	0.7	0.9
IA_4	0.4	0.6	0.9	0.6	0.1
IA_5	0.5	0.6	0.7	0.6	0
IA_6	0.8	0.8	0.7	0.9	0

We are developed an IDM model using *TNP* and *TCP* by selecting a set of factors from $FAVS(\Psi, D)$ chosen by an investor to classify the avenue that suits the best requirements.

Example 5.3 Let the preferences of factors by an investor Mr. Singh are as follows:

(i) High returns (IF2),
(ii) Stable return (IF3) and
(iii) Tax concession (IF5).

Then we have the *TNP* $\Psi(IF2) \otimes \Psi(IF3) \otimes \Psi(IF5)$ as in Table 2 and *TCP* $\Psi(IF2) \oplus \Psi(IF3) \oplus \Psi(IF5)$ as shown in Table 3.

Here IA_3 has the highest MV, i.e. $\mu_{\Psi(IF2)\otimes\Psi(IF3)\otimes\Psi(IF5)}(IA_3) = 0.15$, so the NSP is the best suit for Mr. Singh.

Table 2 TNP-$\Psi(IF2) \otimes \Psi(IF3) \otimes \Psi(IF5)$

V	ψ(IF2)	ψ(IF3)	ψ(IF5)	$\Psi(IF2) \otimes \Psi(IF3) \otimes \Psi(IF5)$
IA_1	0.5	0.45	0.75	0.115
IA_2	0.6	0.35	0.9	0.141
IA_3	0.5	0.45	0.9	0.15
IA_4	0.6	0.9	0.1	0.037
IA_5	0.6	0.7	0	0
IA_6	0.8	0.7	0	0

Table 3 TCP-$\Psi(IF2) \oplus \Psi(IF3) \oplus \Psi(IF5)$

V	ψ(IF2)	ψ(IF3)	ψ(IF5)	$\Psi(IF2) \oplus \Psi(IF3) \oplus \Psi(IF5)$
IA_1	0.5	0.45	0.75	0.536
IA_2	0.6	0.35	0.9	0.58
IA_3	0.5	0.45	0.9	0.575
IA_4	0.6	0.9	0.1	0.524
IA_5	0.6	0.7	0	0.433
IA_6	0.8	0.7	0	0.5

Here IA_2 has the highest MV, i.e., 0.58, so in this case, the PPF is the best suit for Mr. Singh.

Similarly, an investor's preference of investment-avenue may be based on any set of investment-factors that the investor prefers.

Example 5.4 If the preferences of factors by another investor Mrs. Singh are

(i) Interest rates (IF1),
(ii) Stable return (IF3),
(iii) Minimum risk (IF4) and
(iv) Tax concession (IF5).

Then we have the *TNP* $\Psi(IF1) \otimes \Psi(IF3) \otimes \Psi(IF4) \otimes \Psi(IF5)$ as in Table 4 and also *TCP* $\Psi(IF1) \oplus \Psi(IF3) \oplus \Psi(IF4) \oplus \Psi(IF5)$ as shown in Table 5.

Here IA_1 has the highest MV, i.e., 0.182, so recurring deposit is the best suit for the requirement of Mrs. Singh.

Here IA_3 has the greatest MV, i.e., 0.6840, so NSP is the best suit for the requirement of Mrs. Singh.

Advantages 5.6: By employing Algorithm1, we can be able to obtain a smaller number of object options, allowing us to settle the IDM more effortlessly. However, by using our method, we can obtain more comprehensive data, which will help leaders in their decision-making.

Table 4 TNP-$\Psi(IF1) \otimes \Psi(IF3) \otimes \Psi(IF4) \otimes \Psi(IF5)$

V	ψ(IF1)	ψ(IF3)	ψ(IF4)	ψ(IF5)	$\Psi(IF1) \otimes \Psi(IF3)$ $\otimes\Psi(IF4) \otimes \Psi(IF5)$
IA_1	0.85	0.45	0.85	0.75	0.182
IA_2	0.8	0.35	0.8	0.9	0.149
IA_3	0.85	0.45	0.7	0.9	0.179
IA_4	0.4	0.9	0.6	0.1	0.011
IA_5	0.5	0.7	0.6	0	0
IA_6	0.8	0.7	0.9	0	0

Table 5 TCP-$\Psi(IF1) \oplus \Psi(IF3) \oplus \Psi(IF4) \oplus \Psi(IF5)$

V	ψ(IF1)	ψ(IF3)	ψ(IF4)	ψ(IF5)	$\Psi(IF1) \oplus \Psi(IF3)$ $\oplus\Psi(IF4) \oplus \Psi(IF5)$
IA_1	0.85	0.45	0.85	0.75	0.683
IA_2	0.8	0.35	0.8	0.9	0.678
IA_3	0.85	0.45	0.7	0.9	0.684
IA_4	0.4	0.9	0.6	0.1	0.497
IA_5	0.5	0.7	0.6	0	0.45
IA_6	0.8	0.7	0.9	0	0.6

6 Conclusion

In this chapter, we characterize TNP and TCP on *FAVSs* and then applying these products we have implemented a novel machine learning approach to FSS-based IDM, in order to invest in a risky situation. Some numerical examples demonstrate the viability of our proposed machine learning method in practice. We believe that this investigation in this area can be done ahead of time. The approach should be expanded in the future to address relevant issues such as computer science, economics, software engineering, machine learning, and so on.

References

1. Ali, M.I., Feng, F., Liu, X., Min, W.K., Shabir, M.: On some new operations in soft set theory. Comput. Math. Appl. **57**, 1547–1553 (2009)
2. Bhardwaj, N., Sharma, P.: An advanced uncertainty measure using fuzzy soft sets: application to decision-making problems. Big Data Min. Anal. **4**(2), 94–103 (2021)
3. Chinram, R., Hussain, A., Ali, M.I., Mahmood, T.: Some geometric aggregation operators under q-rung orthopair fuzzy soft information with their applications in multi-criteria decision making. IEEE Access **9**, 31975–31993 (2021)
4. Dalkılıç, O.: A novel approach to soft set theory in decision-making under uncertainty. Int. J. Comput Math. (2021). https://doi.org/10.1080/00207160.2020.1868445
5. Dalkılıç, O., Demirtaş, N.: Bipolar fuzzy soft D-metric spaces. Commun. Fac. Sci. Univ. Ankara Ser. A1 Math. Stat. **70**(1), 64–73 (2021)
6. Dalkiliç, O.: On topological structures of virtual fuzzy parametrized fuzzy soft sets. Complex Intell. Syst. (2021). https://doi.org/10.1007/s40747-021-00378-x
7. Deli, I., Çağman, N.: Intuitionistic fuzzy parameterized soft set theory and its decision making. Appl. Soft Comput. **28**, 109–113 (2015)
8. Deng, X., Jiang, W.: D number theory based game-theoretic framework in adversarial decision making under a fuzzy environment. Int. J. Approx. Reason **106**, 194–213 (2019)
9. Feng, F., Jun, Y.B., Liu, X., Li, L.: An adjustable approach to fuzzy soft set based decision making. J. Comp. Appl. Math. **234**, 10–20 (2010)
10. Gao, R., Wu, J.: Filter with its applications in fuzzy soft topological spaces. AIMS Math. **6**(3), 2359–2368 (2021)
11. Jamkhaneh, E.B., Nadarajah, S.: A new generalized intuitionistic fuzzy set. Hacet. J. Math. Stat. **44**(6), 1537–1551 (2015)
12. Kong, Z., Gao, L.Q., Wang, L.F.: Comment on A fuzzy soft set theoretic approach to decision making problems. J. Comp. Appl. Math. **223**, 540–542 (2009)
13. Lathamaheswari, M., Nagarajan, D., Kavikumar, J., et al.: Triangular interval type-2 fuzzy soft set and its application. Complex Intell. Syst. **6**, 531–544 (2020)
14. Li, Z., Liu, X., Dai, J., Chen, J., Fujita, H.: Measures of uncertainty based on gaussian kernel for a fully fuzzy information system. Knowl. Based Syst. **196**(105791), 03 (2020)
15. Maji, P.K., Biswas, R., Roy, A.R.: Fuzzy soft sets. J. Fuzzy Math. **9**, 589–602 (2001)
16. Maji, P.K., Biswas, R., Roy, A.R.: Soft set theory. Comput. Math. Appl. **45**, 555–562 (2003)
17. Maji, P.K., Roy, A.R., Biswas, R.: An application of soft sets in a decision making problem. Comput. Math. Appl. **44**, 1077–1083 (2002)
18. Mandal, P., Ranadive, A.S.: Multi-granulation fuzzy probabilistic rough sets and their corresponding three-way decisions over two universes. Iran. J. Fuzzy Syst. **16**(5), 61–76 (2019)
19. Močkoř, J., Hurtik, P.: Fuzzy soft sets and image processing application. In: Aliev, R.A., Kacprzyk, J., Pedrycz, W., Jamshidi, M., Babanli, M., Sadikoglu, F.M. (eds.) 14th International

Conference on Theory and Application of Fuzzy Systems and Soft Computing – ICAFS-2020. ICAFS 2020. Advances in Intelligent Systems and Computing, vol. 1306. Springer, Cham (2021). https://doi.org/10.1007/978-3-030-64058-3_6
20. Molodtsov, D.: Soft set theory-first results. Comput. Math. Appl. **37**, 19–31 (1999)
21. Mukherjee, A., Das, A.K.: Application of fuzzy soft multi sets in decision-making problems. In: Nagar, A., Mohapatra, D., Chaki, N. (eds.) Proceedings of 3rd International Conference on Advanced Computing, Networking and Informatics. Smart Innovation, Systems and Technologies, vol. 43. Springer, New Delhi (2016). https://doi.org/10.1007/978-81-322-2538-6_3
22. Mukherjee, A., Das, A.K.: Application of interval valued intuitionistic fuzzy soft set in investment decision making. In: 2015 fifth international conference on advances in computing and communications (ICACC), pp. 61–64 (2015).https://doi.org/10.1109/ICACC.2015.37
23. Nihal, T.: On the pasting lemma on a fuzzy soft topological space with mixed structure. Math. Sci. Appl. E-Notes **8**(2), 15–20 (2020)
24. Paik, B., Mondal, S.K.: A distance-similarity method to solve fuzzy sets and fuzzy soft sets based decision-making problems. Soft Comput. **24**, 5217–5229 (2020)
25. Paik, B., Mondal, S.K.: Representation and application of Fuzzy soft sets in type-2 environment. Complex Intell. Syst. (2021). https://doi.org/10.1007/s40747-021-00286-0
26. Peng, X., Li, W.: Algorithms for hesitant fuzzy soft decision making based on revised aggregation operators, WDBA and CODAS. J. Intell. Fuzzy Syst. **36**(6), 6307–6323 (2019)
27. Petchimuthu, S., Garg, H., Kamacı, H., Atagün, A.O.: The mean operators and generalized products of fuzzy soft matrices and their applications in MCGDM. Comput. Appl. Math. **39**(2), 1–32 (2020)
28. Rajput, A.S., Thakur, S.S., Dubey, O.P.: Soft almost ββ-continuity in soft topological spaces. Int. J. Stud. Res. Technol. Manag. **8**(2), 06–14 (2020)
29. Roy, A.R., Maji, P.K.: A fuzzy soft set theoretic approach to decision making problems. J. Comput. Appl. Math. **203**, 412–418 (2007)
30. Shakila, K., Selvi, R.: A note on fuzzy soft Paraopen sets and maps in fuzzy soft topological spaces. J. Math. Comput. Sci. **10**(2), 289–308 (2019)
31. Smarandache, F., Parimala, M., Karthika, M.: A review of fuzzy soft topological spaces, intuitionistic fuzzy soft topological spaces and neutrosophic soft topological spaces. Int. J. Neutrosophic. Sci. **10**(2), 96 (2020)
32. Vimala, J., Reeta, J.A., Ilamathi, V.S.: A study on fuzzy soft cardinality in lattice ordered fuzzy soft group and its application in decision making problems. J. Intell. Fuzzy Syst. **34**(3), 1535–1542 (2018)
33. Wan Mohd, W.R., Abdullah, L., Yusoff, B., Taib, C.M.I.C., Merigo, J.M.: An integrated mcdm model based on pythagorean fuzzy sets for green supplier development program. Malays. J. Math. Sci. **13**, 23–37 (2019)
34. Zadeh, L.A.: Fuzzy sets. Inf. Control **8**, 338–353 (1965)
35. Zhan, J., Alcantud, J.C.R.: A novel type of soft rough covering and its application to multicriteria group decision making. Artif. Intell. Rev. **52**(4), 2381–2410 (2019)
36. Zhu, K., Zhan, J.: Fuzzy parameterized fuzzy soft sets and decision making. Int. J. Mach. Learn. Cybern. (Springer), available as online first article. (2015). https://doi.org/10.1007/s13042-015-0449-z

A Design and Implementation of Ring Oscillator Physically Unclonable Function Using the Xilinx FPGA

Swati Kulkarni, R. M. Vani, and P. V. Hunagund

1 Introduction

What do you consider to be the most significant aspect of security? Banking, the military, data on the internet and networking are all businesses in which security is critical. Embedded security is also used in numerous industries, such as hardware security. Let's start with an explanation of what hardware security entails. Security-related hardware platforms include ASICs, semi-custom, and fully bespoke ICs. FPGAs are getting more popular as a result of advancement and revolution. FPGAs are preferred over full-custom ICs and ASICs because they are reusable and reprogrammable. Researchers are now looking toward SoC-based FPGAs. A built-in high-speed CPU, a large number of FPGA resources, and a shorter time to market distinguish SoC-based FPGAs. The design time is reduced because IPs are pre-defined and pre-verified. To launch a product in a certain amount of time, most designers supply intellectual property (IP). If third-party IP contains a hardware trojan that is damaging to the design, it should be avoided. Due to their increasing demand, FPGAs have become a popular target for piracy [1, 2]. Because of advances in attacking tactics, the secret key used in traditional cryptography is no longer safe. The Cryptographic key is frequently stored in EEPROM or battery backup SRAM. An attacker can utilize side-channel attacks to try to gain the cryptographic key once the memory has been started up. Another disadvantage of keeping the cryptographic key inside the EEPROM or battery-backed SRAM is that it raises the cost by increasing the hardware complexity [3]. That address these issues, we need a promising solution to enable hardware security. The PUF is a blessing in disguise. There is no need to

S. Kulkarni (✉) · P. V. Hunagund
Department of Applied Electronics, Gulbarga University, Gulbarga 585106, India
e-mail: swatikulkarni494@gmail.com

R. M. Vani
University Science Instrumentation Centre, Gulbarga University, Gulbarga 585106, India

S. Mukherjee et al. (eds.), *Intelligent Systems for Social Good*,
Advanced Technologies and Societal Change,
https://doi.org/10.1007/978-981-19-0770-8_13

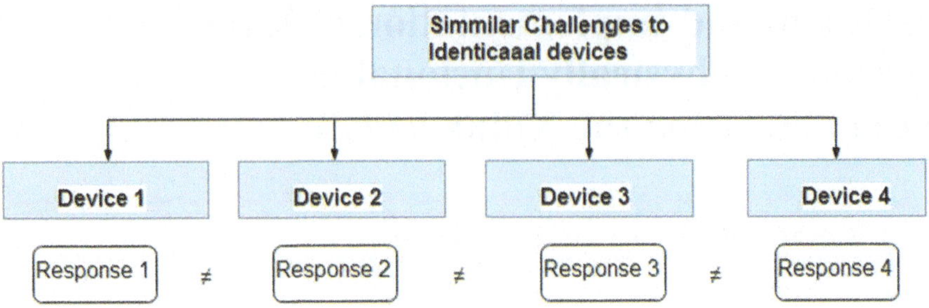

Fig. 1 Uniqueness of PUF (*Source* http://www.ictk.co.kr/servicenproduct/puf)

Fig. 2 Reliability of PUF (*Source* https://cryptography.gmu.edu/research/pufs.php)

save the key in PUF-based encryption, and no one can clone or replicate it. PUF focuses on Integrated Circuits' inherent nanoscale process changes (IC). The power, voltage, and temperature changes of an IC are all elements to consider (PVT). Due to manufacturing variance, even two identical ICs or components can have different propagation delays, which can be uncontrollable. PUF accepts input as a challenge and responds with output. Each time PUF is used, and it creates a different and random response [4].

Even if two identical ICs are applied to the same challenge, their responses will differ (see Fig. 1). This one-of-a-kind response may be considered a device signature. This is one of the most distinctive characteristics of PUF [5]. Figure 2 shows another important aspect of PUF which is its reliability. If a particular challenge is applied to the PUF device, it produces a response. If the same challenge is given to the same device after a few years, it should generate the same response as before [6].

The applied input determines the output of any mathematical or software function; however, the output of PUF is determined not just by the applied input, but also by the device's internal variation. This feature of PUF makes it different than other mathematical functions or software functions. Please keep in mind that the PUF response is not a cryptographic key. The response bits must undergo error repair and detection.

1.1 Related Work

The two types of PUF are silicon-based PUF and non-silicon-based PUF. The first PUF, an optical PUF or non-silicon PUF, was created by authors [7]. Silicon-based PUF comes in two types: delay-based and memory-based. The most prevalent delay-based PUF is an Arbiter PUF (APUF), which is also known as a Strong PUF because it contains the most challenge-response pairings. Weak PUF was the name given to RO-PUF when it was originally introduced to a small number of challenge-response pairs by [8]. Author [9] described a PUF based on a modified Ring oscillator (RO-PUF). Delay PUF assesses devices' propagation delays and develops unique replies by posing challenges to them. The Arbiter PUF is powerful because it has a large number of challenges and response pairings (CRPs). Because it contains a low number of CRPs, RO-PUF is a weak PUF. Routing symmetry is required for APUF, which is challenging to implement on an FPGA [10]. On Arbiter PUF, modeling attacks are more widespread. RO-PUF does not require symmetric routing and maybe developed quite quickly on an FPGA board [11]. We implemented and tested the PUF on FPGAs in this study. ASICs are less flexible than FPGAs. Because FPGAs are preconfigured, FPGA design requires less time to market. In comparison to ASIC fabrication, FPGAs are a low-cost option. These are the main reasons why we chose FPGA technology over ASIC implementation.

In 2012, the Xilinx came with seven series unified architecture FPGA with a new tool, Vivado. The Xilinx Launched 28 nm technology, System on chip (SoC) boards. SoC is a combination of a hard-wired Processing System (PS) and Programmable Logic (PL) that consists of many Configurable Logic Blocks (CLBs) and a complex routing system. Figure 3 shows the device view on the Xilinx Zed Board with PS and PL. PS refers to the black box in the left corner, while PL refers to the rest of

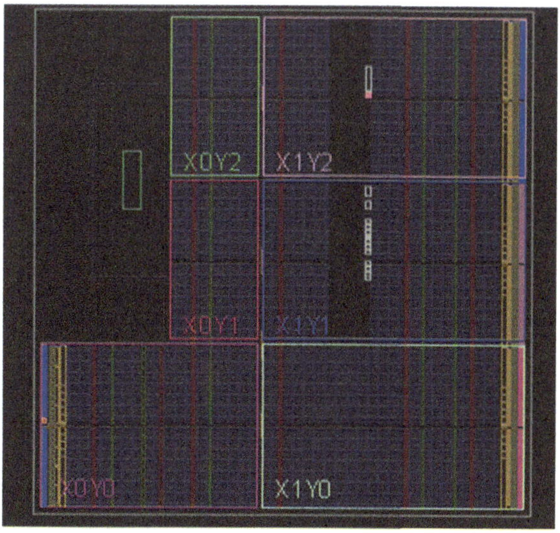

Fig. 3 Device view with PS and PL [12]

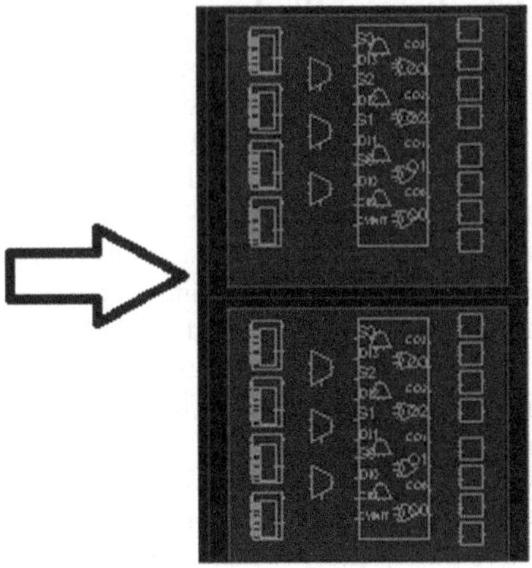

Fig. 4 Single CLB structure

the colorful marking zones. PS and PL are both available on a single chip, which is referred to as an SoC. On the PS side, the Xilinx Zed Board has a dual-core ARM A9 dedicated CPU, and on the PL side, the Artix 7. The CLBs consists of a switch matrix that connects the slices and the outside world. Slice L and slice M are the two slices that make up a CLB shown in Fig. 4. LUTs are used to create logic functions, whereas FFs are used to store data in a single slice. Each CLB on the Xilinx Zynq chip used in this study contains two slices. Four LUTs and eight FFs make up each slice. The SoC-based design can fulfill these requirements. In this chapter, we have implemented RO-PUF and tested it on the Xilinx Zed Board (SoC) [13]. The main focus of this paper is on an area-optimized RO PUF implementation with good reliability. The design was tested for its area usage, speed, uniqueness, uniformity, and reliability. With a brief introduction to PUF, we discussed the Ring oscillator PUF with mathematical delay model in the second section, the implementation of RO PUF on FPGA in the third section, results in section four, and a conclusion in the fourth section.

2 Ring Oscillator

Figure 5 shows a schematic of the Ring oscillator formed using one NAND Gate and three inverters. RO will generate a square waveform. The frequency of the square wave depends on the net delays and propagation delay of each gate [14].

Figure 6 shows the basic block diagram of the 1-bit RO PUF. Ring oscillator PUF design comprises 32 ROs. The first sixteen ROS connected to one of the 16:1 multiplexer another 16 ROs connected to another 16:1 multiplexer. Select lines of multiplexers will decide which RO to be selected. Both multiplexers will use select lines, i.e., 'challenge' input, to choose any two ROs at the same time. Multiplexers

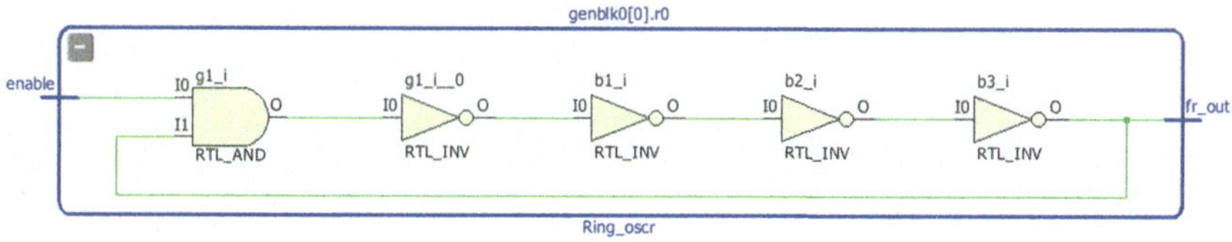

Fig. 5 Ring oscillator [15]

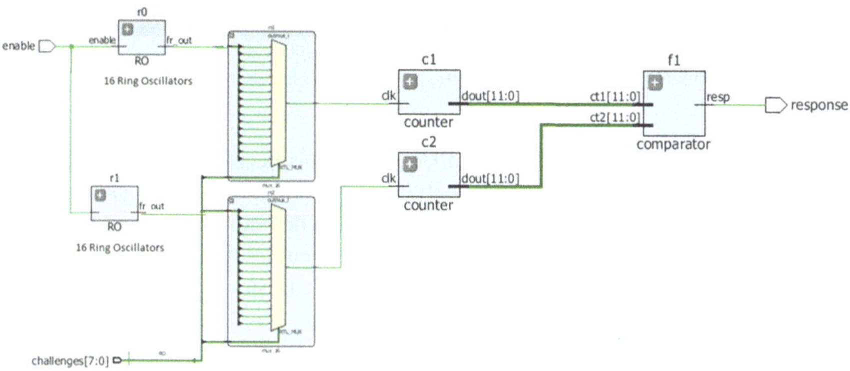

Fig. 6 One-bit RO PUF schematic

connected clock input to two distinct 12-bit counters via the output of multiplexers. The counting will be carried out on every affirmative edge of ROs. The comparator compares the counter output. As previously said, each RO generates frequencies with distinct periods. Even if both counters are linked to the multiplexers, counting may start at the respective incoming positive edge. It causes any one of the counters may get overflows first. If the upper counter gets overflowed first than the lower counter then comparator output will be '1' otherwise '0'. The output of the comparator is a 'response' bit. ROs on two different devices have different frequencies. This difference allows the RO PUF to characterize devices to authenticate them [1, 16].

This is an implementation of the RO-PUF in 8 bits. Figures 7 and 8 demonstrate

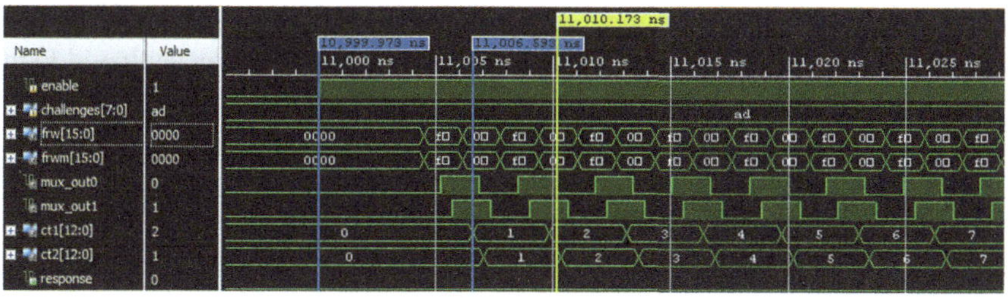

Fig. 7 Post implemented timing simulation (**a**)

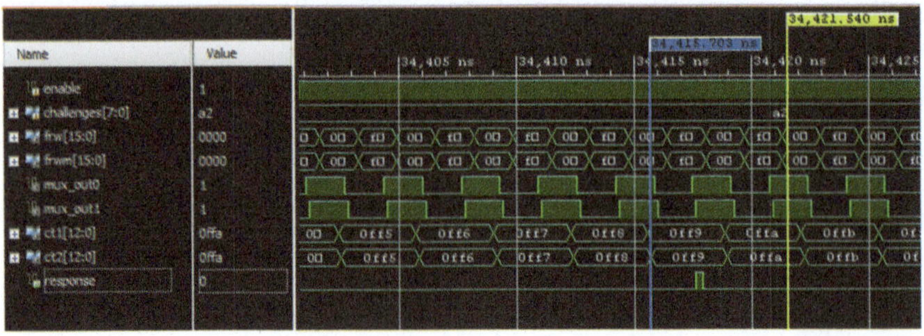

Fig. 8 Post implemented timing simulation (**b**)

the time simulation of a single bit RO PUF after it has been implemented (b). The inputs are Enable and Challenges. The output of each module can be seen here. ROs will produce square waves, as previously indicated. The outputs of Ring Oscillators are frw and frwm. Mux out1 and mux out2 are the mux outputs. The mux outputs are mux out1 and mux out2, and two multiplexers can simultaneously select any two ROs. Two counters use the clock input from these mux outputs as well. Every positive edge of the incoming clock will be counted as well. The outputs of the two counters are referred to as Ct1 and Ct2. Because no counters are overflowing, the response bit remains '0'. If Ct1 reaches its maximum value before Ct2, the response bit will be set to '1'.

In this Fig. 8, we can observe the 'response' bit. As soon as the first counter gets overflows response bit became '1'.

2.1 Mathematical Function of Delay

To understand the RO PUF's operation, we must first know the delay module. According to [5], every path consists of two delay elements, a static delay element and a random delay element. A Static delay element is the delay of a circuit calculated by adding the individual gate and net delays for each path a Random delay element is present due to process variations. Ideally, the PUF output should only be dependent on its process variation. Hence, out of the two components, for a PUF, the random delay component should be the significant factor. As RO-PUF is delay-based PUF, we defined Delay D of a net.

$$D = DS + D\dot{R} \tag{1}$$

where DS is a static delay and DR is a random delay. The Static delay can be determined using the timing tool available in Vivado by applying timing constraints. It will remain the same for the same RTL and Board. But the random delay is delay generated due to internal process variation of devices hence it will be different for each board. Mainly PUF functionality depends on Random delay. The conclusions of the Arbiter, Butterfly and Ring Oscillator PUFs are based on the premise that the symmetrical pathways' static delays cancel out [17].

Let us consider two nets N1 and N2 from the same design which need to compare,

$$D(N1) = Ds1 + DR1 \tag{2}$$

$$D(N2) = Ds2 + DR2 \tag{3}$$

Equations 2 and 3 can be used to express the delay D differences between the two networks. If two nets $N1$ and $N2$ are identical, $Ds1 = Ds2$ and the delay skew D between them can be calculated.

$$\Delta D = D1 - D2 = DR1 - DR2 = \Delta DR \tag{4}$$

The static delay differences DS should tend to zero in an ideal case, and the delay comparison between the two net delays should be a function of the random delay component. However, if $DS1DS2$ (that is, $N1$ and $N2$) are not similar, the delay skew can be determined as follows:

$$\Delta D = Ds1 - Ds2 + DR1 - DR2 = \Delta DS + \Delta DR \tag{5}$$

This equation suggests that the delay difference between two nets is primarily the sum of the difference of the individual components. Even a slight contribution by the static delay component can result in a biased PUF output. If $DS > DR$, then the effect of random variation on the output will be insignificant and the output of PUF depends on static delay rather than random delay. In that case we can say that the output of PUF will be biased [4].

3 Implementation of RO PUF on FPGA

While implementing the RO PUFs on an FPGA, we need to consider the following things. The synthesis tool will always try to optimize a design concerning speed and area. The ROs with unpredictable behavior would either throw a latch warning or be completely optimized out as it serves no purpose from a tool point of view. Additionally, every pair of ROs that is being compared is to be implemented the same. ROs design needs to instantiate many times but the optimization point of view tool will remove the multiple instances. To avoid this, the placement and routing had to be thoroughly constrained. The usage of slices, logic, pins, and routing needs to be done manually to prevent any optimization by the synthesis tool. Ring Oscillator PUFs have the requirement of symmetric routing. We must construct it extremely carefully to avoid biassing of delays and to maintain all ROs in the PUF identical. RO PUF necessitates the placement of ROs by hand. Each slice of the 7series FPGA consists of four LUTs, as we explained previously. Because each RO contains four logic gates, we assigned each LUT to one of them. One RO was implemented per slice. In FPGA, hard macros are advised for manual installation [18].

Hard macros are required to avoid any additional delays caused by the various instantiations required for the implementation. Instead of the entire design being symmetrically placed and routed, the RO PUF simply requires each Ring Oscillator to be similar. Furthermore, utilizing hard macros on Xilinx FPGAs, similar ring oscillator routing is simple [19]. Formerly to perform manual placement and routing, hard macros were employed. In the earlier version of the Xilinx tool, the ISE suite supports an FPGA editor to crate Hard macro. Now the Xilinx Vivado does not support the FPGA editor instead it has a TCL shell. The Xilinx Design Constraint (XDC) constraints are written in TCL scripting. The Tcl script helps to place all the resources at the desired location. Figure 9 shows the device view after the manual placement of the component [15].

Figure 9 shows the TCL script written in the TCL console of the Vivado. The Tool will automatically convert the TCL script into the Xilinx Design Constraints (XDC) shown in Fig. 10. The constraint "**set_property**" is responsible to place the instance into the cell. BEL specifies a specific placement within a Slice for a register or LUT. BEL is generally used with an associated LOC property to specify the exact placement of a register or LUT. A primitive component's LOC indicates where it should be placed (Fig. 11).

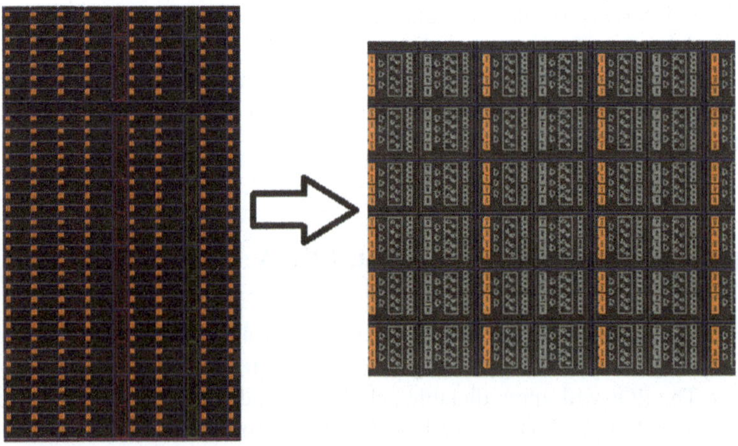

Fig. 9 Device view after manual placement of the components

```
for {set q 0}  {$q <= 15}   {incr q}   {
set current_pos_x [expr $start_pos_x+0]
set current_pos_y  [expr $start_pos_y + $q]

startgroup
place_cell "generate_block_identifier[0].t0/genblk0[$q].r0/fr_out_INST_0" SLICE_X${current_pos_x}Y${current_pos_y}/D6LUT
endgroup
startgroup
place_cell "generate_block_identifier[0].t0/genblk0[$q].r0/n_inferred_i_1" SLICE_X${current_pos_x}Y${current_pos_y}/C6LUT
endgroup
startgroup
place_cell "generate_block_identifier[0].t0/genblk0[$q].r0/n_inferred_i_2" SLICE_X${current_pos_x}Y${current_pos_y}/B6LUT
endgroup
startgroup
place_cell "generate_block_identifier[0].t0/genblk0[$q].r0/n_inferred_i_3" SLICE_X${current_pos_x}Y${current_pos_y}/A6LUT
endgroup
```

Fig. 10 The TCL script

```
set_property BEL D6LUT [get_cells {generate_block_identifier[0].t0/genblk0[0].r0/fr_out_INST_0}]
set_property LOC SLICE_X0Y0 [get_cells {generate_block_identifier[0].t0/genblk0[0].r0/fr_out_INST_0}]
set_property BEL C6LUT [get_cells {generate_block_identifier[0].t0/genblk0[0].r0/n_inferred_i_1}]
set_property LOC SLICE_X0Y0 [get_cells {generate_block_identifier[0].t0/genblk0[0].r0/n_inferred_i_1}]
set_property BEL B6LUT [get_cells {generate_block_identifier[0].t0/genblk0[0].r0/n_inferred_i_2}]
set_property LOC SLICE_X0Y0 [get_cells {generate_block_identifier[0].t0/genblk0[0].r0/n_inferred_i_2}]
set_property BEL A6LUT [get_cells {generate_block_identifier[0].t0/genblk0[0].r0/n_inferred_i_3}]
set property LOC SLICE X0Y0 [get cells {generate block identifier[0].t0/genblk0[0].r0/n inferred i 3}]
```

Fig. 11 The XDC location constraints

4 Experimental Setup and Results

We converted our design to IP for hardware testing. Figure 12 shows the connection diagram. The Xilinx VIO and ILA IPs are interfaced with PUF IP. ILA and VIO are used for Hardware Debugging. The LogiCORE IP Virtual Input/Output (VIO) is a programmable IP core for real-time monitoring of the design's internal signals.

ILA is used to monitor internal programmable logic signals and ports for post-analysis. Using VIO and ILA we are applying the input to our PUF design and observing the outputs. Figures 13, 14, 15, and 16 show the design's real-time output in terms of ILA output waveform for different combinations of CRP.

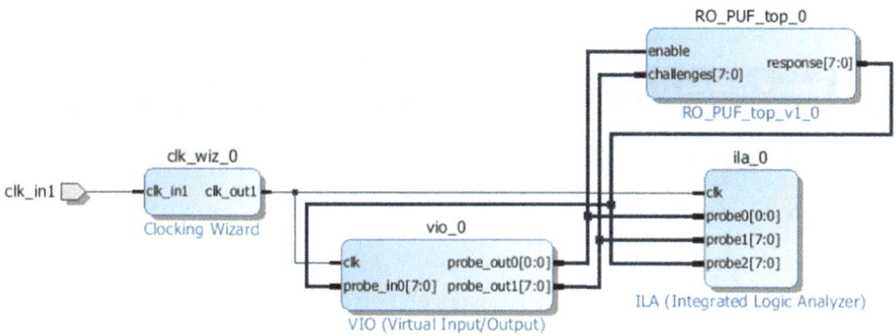

Fig. 12 Block diagram for Hardware debugging

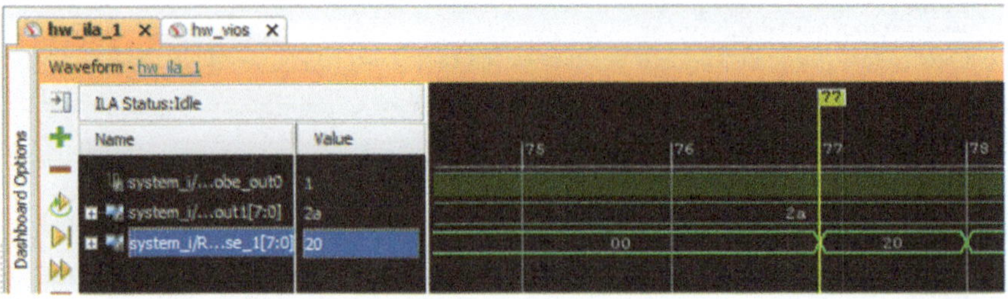

Fig. 13 ILA output waveform (**a**)

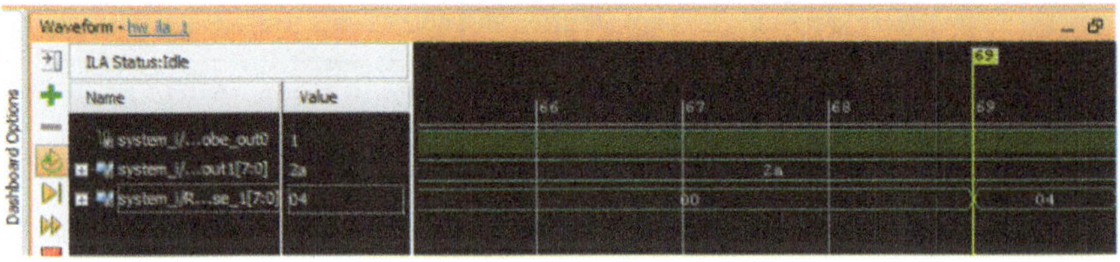

Fig. 14 ILA output waveform (**b**)

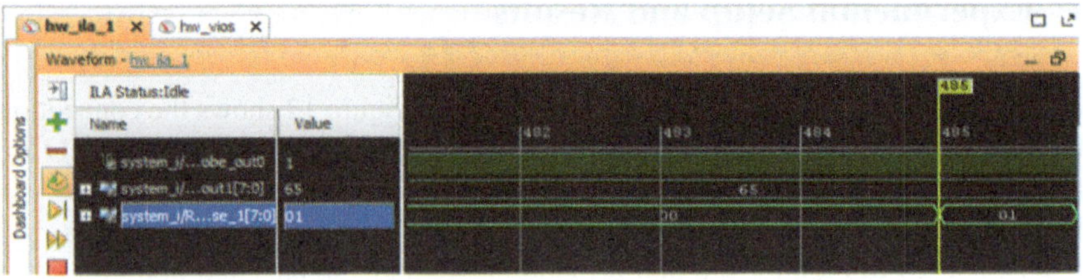

Fig. 15 ILA output waveform (**c**)

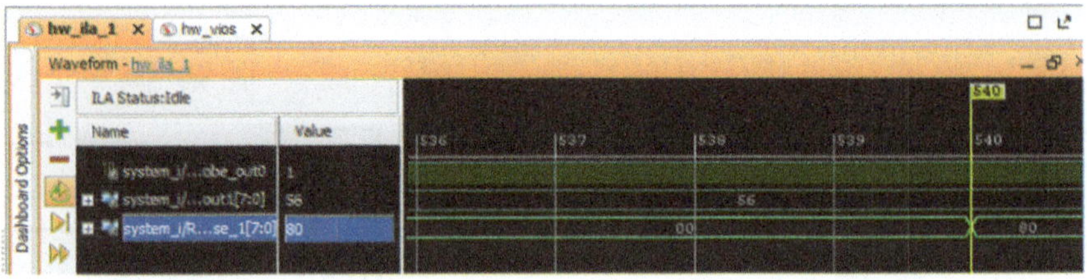

Fig. 16 ILA output waveform (**d**)

We have tested the same RO-PUF design on two different Zed Board FPGA. Here we can observe that even for two similar challenges we are getting different response bits on two different boards. This shows the reliability of the PUF design. Similarly, we got various answer bits when we applied the different challenges to the same

Table 1 We have verified the output for the following cases

Sr. nos.	Criteria	Observations
Case I	Same design on two different boards (inter) with same challenges	We observed unique and random response bits. It proves the uniqueness of PUF
Case II	Same design, same board but different clock region (intra) with same challenges	We observed unique and random response bits
Case III	Same design, Same board, same region but different day and time with same challenges	We observed same response bits again and again. It proves the reliability of PUF
Case IV	Same design, same board, same region but same day and time with multiple challenges	We observed unique and random response bits

Resource	Utilization	Available	Utilization %
LUT	1344	53200	2.53
FF	192	106400	0.18
IO	17	200	8.50

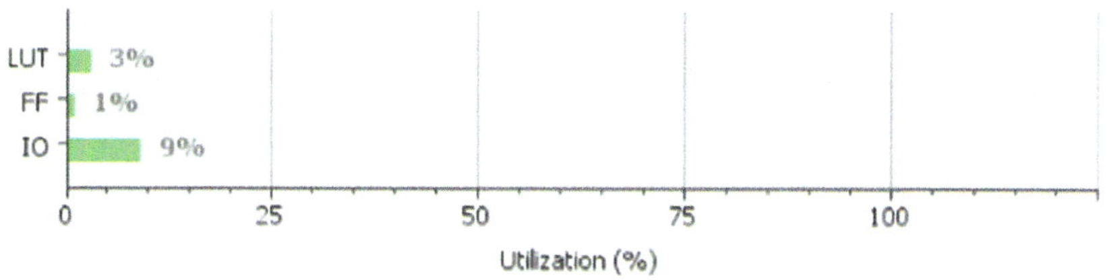

Fig. 17 Utilization summary

design and board. We have presented two challenges at random. We have confirmed that we are getting the unique answer bits for all of the tasks.

Figure 17 shows the utilization summary of the resource. The design uses less than 3% of the total FPGA Logic resources and less than 10% of IO resources.

The design's power analysis is shown in Fig. 18. The total onchip power, junction temperature, and thermal margin are all defined via power analysis. The implemented system logic consumes 61% of the total power available.

5 Conclusion

Hardware-oriented security is an upcoming field in the electronics industry. Today hardware designers are paying more focus on hardware security. Many different PUF designs are getting designed by the researchers. The RO PUF architecture is straightforward, but the FPGA implementation proved difficult. We did so till the RO

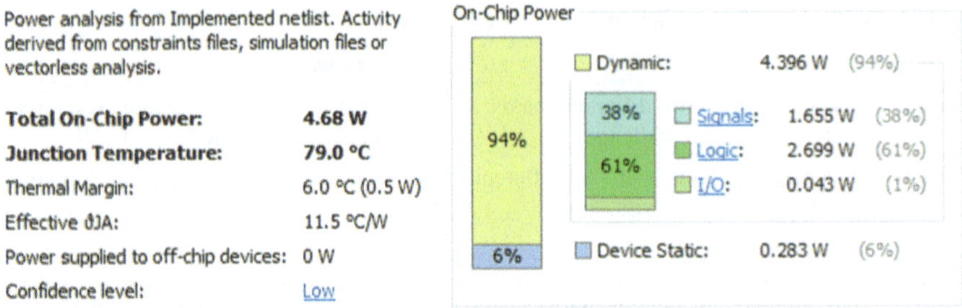

Fig. 18 Power report

PUF on FPGA was successfully implemented. We verified the uniqueness property and reliability property of PUF. However, PUF output is noisy, the generated response needs to filter out using an error correction algorithm. After that, the FPGA board will connect to the computer for more Hamming distance and Entropy research.

References

1. Kumar, M.A., Bhakthavatchalu, R.: FPGA based delay PUF implementation for security applications. In: IEEE International Conference on Technological Advancements in Power and Energy (TAP Energy). Kollam, India (2017)
2. Yin, C., Qu, G., Zhou, Q.: Design and implementation of a Group-based RO PUF. In: Automation & Test in Europe Conference & Exhibition. Grenoble, France (2013)
3. Lim, D., Lee, J.W., Gassend, B., Suh, G.E., van Dijk, M., Devadas, S.: Extracting secret keys from integrated circuits. IEEE Trans. Very Large Scale Integr. (VLSI) Syst. **13**, 1200–1205 (2005)
4. Bernard, F., Fischer, V., Costea, C., Fouquet, R.: Implementation of ring-oscillators-based physical unclonable functions with independent bits in the response. Int. J. Reconfig. Comput, **2012**, 13 (2012)
5. Suh, G.E., Devadas, S.: Physical unclonable functions for device authentication and secret key generation. In: 44th ACM/IEEE Design Automation Conference. San Diego, CA (2007)
6. Handschuh, H., Schrijen, G.-J., Tuyls, P.: Hardware intrinsic security from physically unclonable functions. In: Sadeghi, A.-R., Naccache, D. (eds.) Towards Hardware-Intrinsic Security, Information Security and Cryptography, pp. 39–53. Springer, Berlin Heidelberg (2011). https://doi.org/10.1007/978-3-642-14452-3_2
7. Wallrabenstein, J.R.: Practical and secure IoT device authentication using physical unclonable functions. In: 2016 IEEE 4th International Conference on Future Internet of Things and Cloud (FiCloud), pp. 99–106. Vienna (2016)
8. Kodýtek F., Lórencz, R.: A design of ring oscillator based PUF on FPGA. In: IEEE 18th International Symposium on Design and Diagnostics of Electronic Circuits & Systems, pp. 37–42. IEEE (2015)
9. Bhargava, M., Mai K.: An efficient reliable PUF-based cryptographic key generator in 65nm CMOS design. In: Automation & Test: in Europe Conference & Exhibition. Dresden (2014)
10. Alkatheiri, M.S., Zhuang, Y., Korobkov, M., Sangi, A.R.: An experimental study of the state-of-the-art PUFs implemented on FPGAs. In: IEEE Conference on Dependable and Secure Computing. Taipei, Taiwan (2017)

11. Kokila, J., Ramasubramanian, N.: Enhanced authentication using hybrid PUF with FSM for protecting IPs of SoC FPGAs. J. Electron. Test. **35**, 543–558. (2019)
12. The Xilinx Xilinx Vivado Design Suite User Guide Synthesis UG901 (v2019.2) - January 27, 2020
13. Pappu, R., Recht, S.B., Taylor, J., Gershenfeld, N.: Physical one-way functions. Sci. J. 5589. **297**, 2026–2030
14. Morozov, S., Maiti, A., Schaumont, P.: A comparative analysis of delay based PUF implementations on FPGA: IACR. Cryptol. ePrint Arch. 629 (2009)
15. Kulkarni, S., Vani, R.M., Hunagund, P.V.: FPGA based hardware security for edge devices in internet of things. In: 5th International Conference on Communication and Electronics Systems (ICCES), pp. 1133–1138. IEEE, Coimbatore, India (2020)
16. Maiti, A., Schaumont, P.: Improved ring oscillator PUF: an FPGA-friendly secure primitive. J. Cryptogr. **24**, 375–397 (2011)
17. Tajik, S., Dietz, E., Frohmann, S., Seifert, J.-P., Nedospasov, D., Helfmeier, C., Boit, C., Dittrich, H.: Physical characterization of arbiter PUFs. Springer, Berlin Heidelberg (2014)
18. Gehrer, S., Leger, S., Sigl, G.: Aging effects on ring-oscillator-based physical unclonable functions on FPGAs. In: 2015 International Conference on ReConfigurable Computing and FPGAs (ReConFigure). Mexico City (2015)
19. Xin, X., Kaps, J.-P., Gaj, K.: A configurable ring-oscillator-based PUF for Xilinx FPGAs. In: 4th Euromicro Conference on Digital System Design—Oulu, pp. 651–657 (2011)
20. Lee, J.W., Lim, D., Gassend, B., Suh, G.E., van Dijk, M., Devadas, S.: A technique to build a secret key in integrated circuits for identification and authentication applications. In: Symposium on VLSI Circuits. Digest of Technical Papers (IEEE Cat. No.04CH37525), pp. 176–179. IEEE, Honolulu, HI, USA (2004)
21. Eiroa, S., Baturone, I.: An analysis of ring oscillator PUF behavior on FPGAs. In: International Conference on Field-Programmable Technology, New Delhi, India, pp. 1–4 (2011)

IOT Based Automated Fish Tank

Yashika Gaidhani, Manisha Waje, Monica Kalbande, and Tejswini Panse

1 Introduction

From last 20 years at a steady pace pet ownership has been increasing. Nowadays freshwater fish is the most popular pet. Fish aquarium maintenance is a very difficult task. Whenever you have to feed, or you have to clean up your fish pot/tank, you have to do a lot of things. The fishes required to be fed two times a day which causes the owner to give food to the fish manually which makes the duty of conserving an aquarium very difficult. At periods when the owner is not available, he cannot feed the fish and has no control over the aquarium. So for a mini aquarium tank the important parameters should be intimately observed and proper actions should be rapidly taken at the time of dangerous situations. According to several studies for maintaining healthy fish, small aquarium tanks are not suitable. General suggestion for aquarium tanks is five gallons which lead to the less impact of mistake on fish. YI-BING LIN proposed the Fish Tank system that utilizes actuators that drive with the aquarium sensors and also implemented an intelligent fish feeding system in which the fish is not either over fed nor under fed, and at the same time, the fish owner can enjoy watching fish feeding remotely [1]. Aaron Don M. Africa proposed system which is used to measure, monitor, and regulate all the important parameter factors related to aquarium system. In real-time temperature and pH measurement was relatively easy as compared to DO sensor [2]. Jui-Ho Chen proposed a system that uses mobile device to monitor the fish farm Environmental Data such as PH value, Temperature, water level sensing, dissolved oxygen. It uses Zigbee module for Wi-Fi. Noor proposed automatic fish feeder system based on PIC microcontroller applications. It consists

Y. Gaidhani (✉) · M. Kalbande · T. Panse
Yeshwantrao Chavan College of Engineering, Nagpur, India
e-mail: shweta29gaidhani@gmail.com

M. Waje
G.H.R,C, Wagholi, Pune, India

S. Mukherjee et al. (eds.), *Intelligent Systems for Social Good*,
Advanced Technologies and Societal Change,
https://doi.org/10.1007/978-981-19-0770-8_14

of pellet stand storage, DC motor, former, and microcontroller. Controlling of pallet is done by DC motor the rotation speed [3].

The system with which we came up is an Automatic Fish Tank. In this chapter this system is a programmed system to take think about fishes. With this system fish aquarium manual maintenance will change with its automated functions. Functions performed by the system are as follows: This system will observe the physical changes in the water and with required changes will maintain as per the ideal conditions. The system will perform all the steps automatically like turbidity of water, monitor lighting, pH detection, temperature control, feeding and water filtration. Overfeeding is the number one problem arises due to fish owners, as unused food will contaminate the water inside the tank due to which the turbidity of water changes, the pH of water changes and the dissolved oxygen level might get low. Overfeeding would also have a harmful effect on the health of fishes, since they do not know how much food to eat and end up eating excess and this would just shorten the life of fishes and their behaviour would be altered. So, the feeding would be done by an automatic feeder that would always drop the precise portion of food time to time. The other parameters like temperature are controlled by the system. Survey indicates that higher temperatures within the optimal temperature range from 25 to 27 °C for the species typically lead to stronger and healthier fish. Rather than stable high or low temperatures, extreme changes in temperature are more dangerous to fish. The pH sensor is used to measure acidity of water which is created by dissolved carbon dioxide the ideal pH range for fishes is 6.8–7.6, even if certain fish may need higher or lower levels. A fish tank needs oxygen to support the livestock. In the water combination of decreased oxygen concentration and eminent carbon dioxide concentration leads to suffocation. The oxygen necessities change depending on the weight and the type of fish. The oxygen necessities will be different according to the type and the weight of fish. Therefore, to avoid oxygen depletion, oxygen levels should be higher than 2–4 mg/L hence DO sensor would monitor the DO level in the tank [5]. As the amount of suspended solid increases, the water muddiness or vagueness increases. Turbidity is the measure of relative clarity of water and waters visibility is controlled by this feature. The ideal range of turbidity is from 1 TO 5 NTU (Nephelometric Turbidity unit) 1 mg/L and is equivalent to 3NTU. The turbidity levels depend upon the type of fish so the range may differ accordingly. Turbidity sensor being used here helps to monitor the turbidity of the water, if the turbidity level is above the desired level, then the water filter controls the cloudiness of the water in the tank.

2 Factors for Water Quality

2.1 pH

pH of the water is affected by fish waste, topping off the water, water hardness, and water evaporation. Drastic changes in the water cause changes in blood pH, which leads to tension and loss. The pH sensor will automatically monitor the pH of the water regularly and indicate it on the LCD screen. Suitable range of pH is 6.8–7.6.

2.2 *Temperature*

As fish are aquatic vertebrates, water temperature greatly affects their immunity and metabolism. The fish immune response is suppressed by decrease in water temperature. Sensor monitors the temperature of the water and displays the value on LCD screen. The Peltiers are used to manage the temperature of the water inside the tank, and it is used for both cooling and heating according to the temperature of the tank.

2.3 *Turbidity*

Turbidity sensor being used here helps to monitor the turbidity of the water, if the turbidity level is above the desired level, then the water filter controls the cloudiness of the water in the tank. The ideal range of turbidity is from 1 TO 5 NTU (Nephelometric Turbidity unit).

2.4 *Dissolved Oxygen*

Atmospheric air and photosynthetic planktons are principal source of oxygen in water. Due to low solubility of oxygen in water, obtaining adequate oxygen is a larger problem for aquatic organisms than terrestrial ones. Solubility of oxygen decreases with factors like low atmospheric pressure, high concentration of submerged plants, high humidity, increase in salinity, increase in temperature, plankton blooms. Reduction of oxygen in water cause starvation, reduced growth, more fish mortality and poor feeding of fish.

3 Components of Proposed System

The proposed systems main components with Arduino controller are shown in block diagram in Fig. 1.

The main objective of this system is to construct and design an automatic aquarium for the people who cannot take care of their fish aquarium daily and reduce the manual factor as much as possible. The system is extremely easy to operate and can be modified according to future requirement. Whenever there is a change in the aquatic parameter of aquarium such as turbidity level, pH level, dissolved oxygen

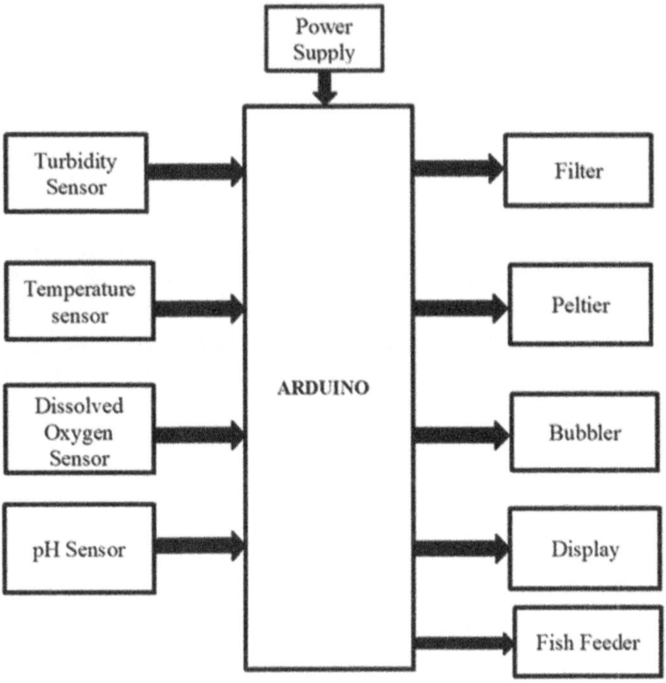

Fig. 1 Block diagram of proposed system

level or in temperature, the sensor will sense these changes, and these changes will then process by the Controller (Arduino Uno). Commands are send by Controller to the sensors, where to sustain the ideal conditions the output part will be observed working.

There will be a pH level sensor, temperature sensor, water level sensor, turbidity sensor, feeder, Peltier and LCD. All the sensors will be interfaced with the controller. If something changes, the controller will start functioning to get back to the ideal state. The normal temperature range of fresh water aquarium is considered to be 25–32 °C. If water temperature of aquarium exceeds, the Peltier will start cooling until the desired temperature is not achieved. If water temperature of aquarium goes down, the Peltier will heat up until the temperature does not reach to the normal temperature. After every 12 h the controller would turn on the feeder for feeding purpose. If the turbidity level has increased beyond the normal range, then controller will start the filter. The value of pH sensor will be display on LCD. The main components of the system are as follows.

3.1 Controller

In this system controlling operations are performed by two Arduino Uno's. These controllers are having 14 digital I/O pins and 6 analogue input pins. Each microcontroller is used for analysing, controlling and processing the input signals from sensors. The Arduino Software (IDE) and Arduino programming language (based on Wiring) are required for writing instruction, based on processing [4].

3.2 pH Sensor

The working pH sensor is used with Arduino controller to find out the alkalinity or acidity of a solution, and it is used with Ardino uno. pH sensor measuring range is 0–14 pH.

3.3 DO Sensor

Analog DO sensor is used to determine the dissolved oxygen in water which range from 0.01 to + 35.99 mg/L. In the tank as dissolved oxygen value goes down the microcontroller boosts the air pump (via relay) to supply tank with the oxygen.

3.4 Temperature Sensor

Temperature sensor is used to measure the temperature of water and maintain it within the required range. The required range is from 25 to −28 °C.

Flowchart for operation of the system is shown in Fig. 2.

If the pH of water is too high or too low, it will effect aquatic ecosystem of an aquarium and results in demise of fish heavy metals and toxicity of chemicals and solubility in the water is affected by pH.

The majority of aquatic creatures preferred a pH range of 6.5–9.0, however some of aquatic creature live in water with pH levels outside preferred range. Same goes for temperature and turbidity.

Aquarium fishes are very sensitivity to temperature changes, for keeping your fish healthy and comfortable good temperature control is essential. The typical range lies between 25 and 30 °C. For controlling the temperature, we are using two peltiers, one is used to for cooling (if temperature is above the desired range) and the other is used for heating (if temperature is below the desired range). If a voltage is positioned across a Peltier element, one side is cooled and the reverse side simultaneously heats up [6]. In Peltier element the hot and cold sides can be swapped by changing the

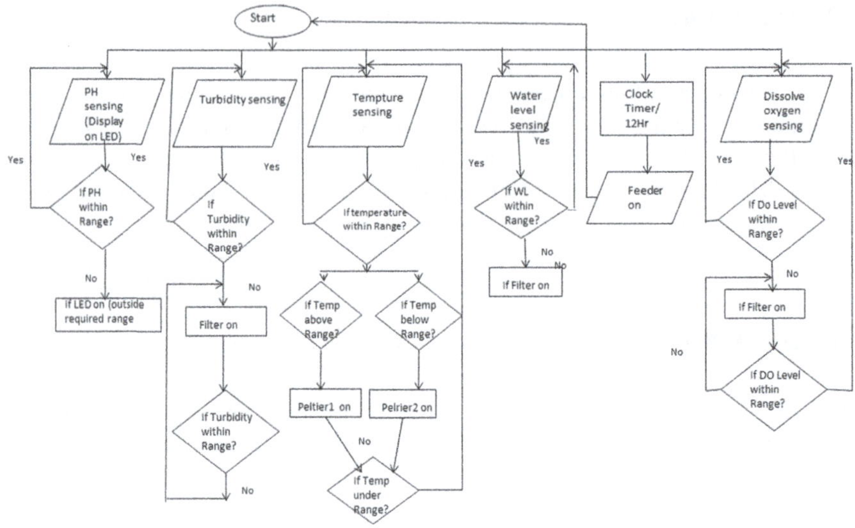

Fig. 2 Flowchart for operation

polarity of the supply voltage. The changes in temperature will be displayed on the LCD.

Inside the tank visibility is limited by cloudy aquarium water. Turbidity may appear in the form of milky-grey water, or as suspended particles visible to the naked eye, or as a greenish stain in the water. So, to avoid turbidity, we are using Turbidity sensor and Filter. If the turbidity is not within the range 70–140 ppm then filter is turned on and after filtration again the turbidity is checked for the required range, process ends when the ideal value is obtained. Mechanical design of system is shown in Fig. 3.

Fig. 3 Mechanical design of system

4 Some Common Mistakes

4.1 Overfeeding Fish and Invertebrates

Adequate supply of food to fish is important but uneaten food just lays on the bottom of the tank, creating ammonia nitrite and nitrate and overloading the biological filter, which ultimately increases the alkalinity of water. The fish food contains high protein that is fed regularly which leads to liver damage, affects metabolism and immunity.

4.2 Inadequate Filtration

There are a number of methods available for filtration but for the tank if proper filter selection is unavailable lead to a wide variety of difficulties. Whether it be biological, chemical, or mechanical, it is better to have more filtration for the size of aquarium. The short of good quality water flow all over the system can cause problems such as the build-up of nuisance algae with low DO (dissolved oxygen).

4.3 Altering Temperature

For the fish, the body temperature is the same as the water temperature which they live in because they are cold blooded animals. The water temperature affects the fish metabolic rate, and it is closely linked to it. Change in water temperature may also increase the mortality rate.

4.4 Replacing All the Water

Changing water frequently or changing the entire water present in the tank and using tap water causes destroy of the beneficial bacteria and ecosystem which has built up in tank. Deficiency of these beneficial bacteria lead to a spike in ammonia which is toxic to fish.

5 Results and Analysis

In the present world to reduce work pressure of human, everything is moving towards automation. Therefore in this chapter we proposed automated fish monitoring system that enables easy control over the several issues of aquarium such as temperature

Table 1 Additive complexity of CFFT

Water Criteria	Acceptable range	Desirable range	Critical
Temperature	25–30 (°C)	25–32 (°C)	<15, >35 (°C)
pH	7–9.5	6.5–9	<4, >11
Alkalinity	50–200 (mg L^{-1})	25–100 (mg L^{-1})	<20, >300 (mg L^{-1})
Hardness	>20 (mg L^{-1})	75–150 (mg L^{-1})	<20, >300 (mg L^{-1})
Turbidity		70–150 (ppm)	55, >250 (ppm)
Water color	Pale to light green	Light green to light brown	Clear water, Dark green & Brown
Dissolved Oxygen	3–5 (mg L^{-1})	5 (mg L^{-1})	<5, >8 (mg L^{-1})
Ammonia	0–0.05 (mg L^{-1})	0– <0.025 (mg L^{-1})	>0.3(mg L^{-1})
Nitrate	0–100 (mg L^{-1})	0.1–4.5 (mg L^{-1})	>100, <0.01 (mg L^{-1})
Nitrite	0.02–2 (mg L^{-1})	< 0.02 (mg L^{-1})	>0.2 (mg L^{-1})
H_2S	0–0.02 (mg L^{-1})	0.002	Any detectable level
CO_2	0–10 (mg L^{-1}	<5, 5–8 (mg L^{-1})	>12 (mg L^{-1})

variations, feeding schedule, turbidity level, dissolve oxygen level which contains sensors to collect the information mandatory for desired actuation using Arduino. The parameters like temperature, turbidity and pH will be displayed on LCD. Every parameter has a precise range in which fishes can live longer. The acceptable range and the ideal ranges are given in Table 1.

6 Conclusion

The main part of this chapter is the combination of actuators and aquarium sensors. This chapter is started with aim to achieve the simple looking task of designing an Automated Fish tank. As time pass away we experience that this was not an easy task in terms of interfacing the sensors and feeder. The critical issue is money required because for up gradation of this project much more stuff up gradation need to be done. As electronics engineering is concerned this chapter build our practical implementation knowledge. Having an Automated Fish tank will save our time and no need to worried for our fish and their aquariums for long time.

References

1. Lin, Y.-B., Tseng, H.-C.: Fish talk: an Iot based mini Aquarium system. IEEE J. Mag. IEEE Access **7**, 35457–35469 (2018)
2. Africa, A.D.M., Aguilar, J.C.C.A., Lim, C.M.S., Pacheco, P.A.A., Rodrin, S.E.C.: Automated aquaculture system that regulates pH, temperature and ammonia. In: IEEE 9th International

Conference on Humanoid, Nanotechnology, Information Technology, Communication and Control, Environment and Management (HNICEM). IEEE Conferences (2017)
3. Chen, J.-H., Sung, W.-T., Lin, G.-Y.: Automatic Aquarium Monitoring for the fish farm Aquaculture environment. In: 2015 IEEE International Conference on Systems, Man and Cybernetics. IEEE Conferences (2015)
4. Mohapatra, B.C., Bikash, S., Sharma, K.K., Majhi, D.: Development and testing of demand feeder for carp feeding in outdoor culture system. Agric. Eng. Int. CIGR EJ. **10** (2009)
5. Noor, M.Z.H., Hussian, A.K., Saaid, M.F., Ali, M.S.A., Zolkapli, M.: The design and development of automatic fish feeder system using PIC microcontroller control and system. In: Graduate Research Colloquium (ICSGRC 2012). IEEE (2012)
6. Prangchumpol, D.: A model of mobile application for automatic fish feeder aquariums system. Int. J. Model. Optim. **8**(5), (2018)

Cloud Based Examination Hall Authentication System Using Fingerprint Module

Monica Kalbande, Yashika Gaidhani, Tejswini Panse, and Mahadev Mahajan

1 Introduction

Nowadays, Personal identification is more important for electronically interconnected society [1–4]. Authentication of a candidate is a main task in all types of competitive examination. Thousands of students appeared every year for entrance exams like UPSC, NEET, GATE, JEE, CAT, etc. in India. Manual verification of the identity cards like PAN CARD, ADHAAR CARD, etc. are not safe adequate to promise authenticity [2]. The most promising solution for authentication is biometric processing based on fingerprint or iris verification based system. Fingerprint for authentication has been adopted for data analysis as compare to iris verification due to cost factor. Also the system required less power and very easy to implement [5]. The advantage of fingerprint sensor having unlimited capacity of conversion each fingerprint to an image and latterly it processed in the database. The main objective of this system is to eliminate any form impression during exam by employing a more secured means of fingerprint biometrics. The practice of formal written examinations is used since a long time to test the candidates or students regularly in the universities, schools, and other educational institutions. Moreover, many competitive examinations are increasingly employed for the selection of recruitments in the civil services, some professions and the posts in industries. The proposed system helps in eliminating the impersonation occurring during any large-scale examination and prevents any falsification that can happen in an examination. It will also save the tedious paper work that needs to be done during the examination, thus saving the resources and efforts for tracking of the candidate's attendance especially in exams where larger

M. Kalbande (✉) · Y. Gaidhani · T. Panse
Yeshwantrao Chavan College of Engineering, Nagpur, India
e-mail: monicakalbande@gmail.com

M. Mahajan
Jhulelal Institute of Technology, Nagpur, India

S. Mukherjee et al. (eds.), *Intelligent Systems for Social Good*,
Advanced Technologies and Societal Change,
https://doi.org/10.1007/978-981-19-0770-8_15

number of candidates is there. This system uses a fingerprint biometric parameter to ensure that only registered candidate is allowed into the examination hall. Thus, eliminating the impersonation occurring during any examination. The system consists of a portable device which allow authentication and attendance taking through fingerprint scan and thus sends the data to the examination server which can be accessed by faculty members as well and notifies the candidate if he/she was absent for the exam. This system can be used mainly in University Examinations, Competitive Examinations, Certification Programs, Laboratories, Educational institutes, Industries and Offices.

2 Related Work

Abdullah Alshbtat, Mohammad Alfraheed, Nabeel Zanoon designed a fingerprint based system for student's examination authentication using image processing with the application of organized algorithm. The analysis of system done with the help of uploading an Adafruit fingerprint library to the Audruino Uno card. This work provide an authentication processor students while entering the classroom. They verified registration of student for many subjects with accuracy of 98%[6]. Aayush Shrestha, Tanisha Rakshit et al., presents a system for verification of student's registration and authentication using the fingerprint sensor and Atmega328 microcontroller. They also measures the body temperature of the student with LM35 temperature sensor. After that the complete data of the students is transferred to a cloud server where it has been monitored and analyzed. If the student's body temperature is higher than normal body temperature then an email as an alert is sent to the authorities[7]. AbhishekSengupta, Nadar Prince, Ms. Keerthi Unni developed an (IoT) based attendance system for a class room. The finger print module is used for student verification. The student's data is previously saved in the finger print module. The attendance of student is pushed to the web server after verifying finger print pattern. The Wi-Fi Modules used to send the student verification ID and device ID to the web-server and accordingly the MySQL database is updated [8]. Yu Maw Win, Saw Aung Nyein designed a system to automate marking and updating of an attendance in the database of computer. The fingerprint module is used to store Student's fingerprints. Serial communication to the main computer is done with Node MCU. The stored data can be easily available for future usage [9]. Moth Myint Thein, HlaMyoTun, Chaw MyatNwe developed an automatic system with the help of MS Visual studio 2012, MS SQL Server 2012, RFID and finger print module. This system can be used by admin or teacher and student. The admin can update teacher's and student's profiles. They also store details to the database and can access any time the teacher's and student's details [10].

3 Proposed System

The Exam hall Authentication System consists of a power supply section which gives regulated 5 V and 12 V. This is directly applied to the Arduino controller. A Fingerprint Sensor Module (R305), Wi-Fi Module (ESP8266) and LCD Display (16*2) are interfaced with the Arduino controller. A switch is provided for registration and scanning of fingerprint. Software uses: MC programming language embedded C, Framework:.NET 4.0, Front End: ASP. NET Win Forms, Back End: Microsoft SQL Server 2005.The fingerprint module uses a software named as SFGdemo V2.0 to communicate with a pc for the registration, storage and deletion of a fingerprint over the sensor module's storage chip.

3.1 System Level Architecture

The system level architecture is shown in Fig. 1, the block diagram of the Exam attendance authentication system. It consists power supply that regulates 5 V and 12 V, Arduino Uno controller, Fingerprint Sensor Module (R305), Wi-Fi Module (ESP8266) and LCD Display (16*2). The biometric samples of students finger-prints are already collected by using fingerprint module. Before using the system for authentication, the user must enroll their biometric sample and a number assigned

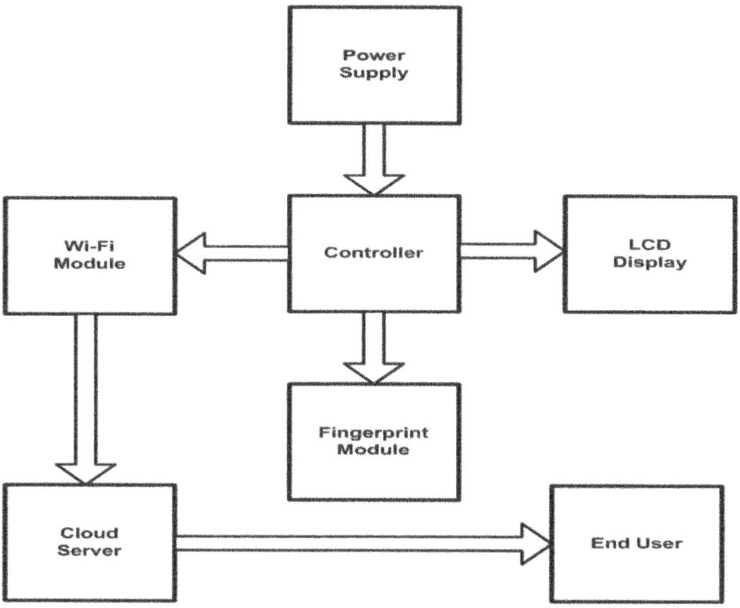

Fig. 1 Block diagram of Exam attendance authentication system

to him/her using the keypad. In the verification mode the system verifies the student by scanning the fingerprint and compares with the stored fingerprints. If the image is registered, it prints outs the individual identification no for confirmation. When a wrong finger is placed on the module, it scans the image, sends to the fingerprint module which try to match the image in its memory. If image is not found, a massage "NOT REGISTERED" is displayed on the screen.

3.2 System Level Operation

The system level operation is described with the flowchart shown in Fig. 2. First we have to provide power supply to the system that will display "FINGERPRINT BASED EXAM HALL AUTHENTICATION". After few seconds, the statement "PUT YOUR FINGER" will pop up. This will show that the fingerprint model scanner is ready for scanning. After scanning the finger, the system may display the statement

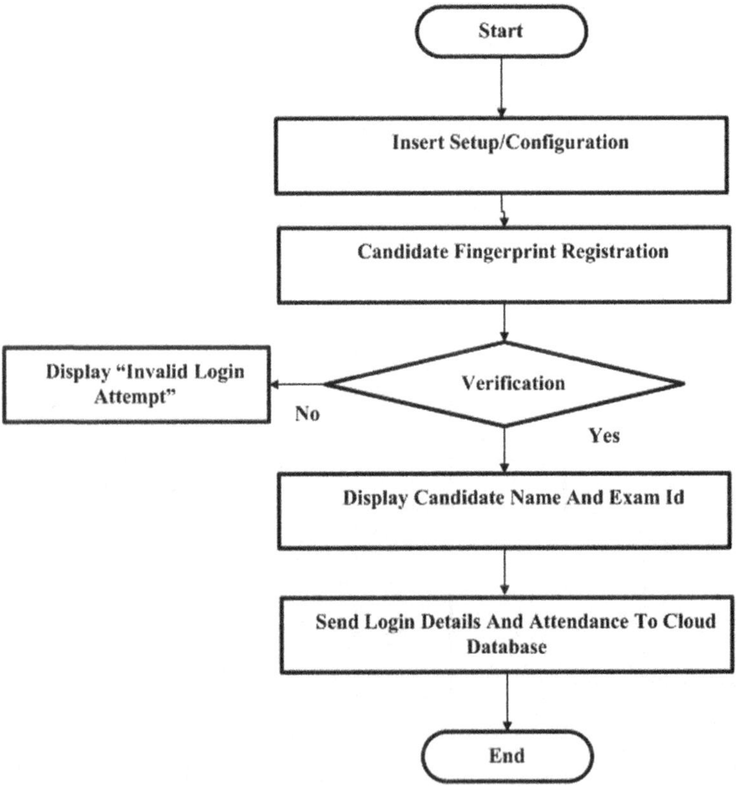

Fig. 2 Flowchart of system level operation for validation of the candidate

"IDENTITY VERIFIED" or it may display "INVALID LOGIN ATTEMPT", if the fingerprint is not verified. The system will further send the login details to the server database from where it can be fetched anytime, anywhere when it will be required.

4 Working Methodology

The working principle of this project is divided into the various stages which are as follows:

Registration mode: The first step in our authentication device is collection of fingerprints of the registered candidates. The fingerprint sensor module is used to register the fingerprints. A particular number is assigned to each candidate using the keypad. It may be called as Registration Number. The Registration Number assigned to each candidate is automatically stored in the fingerprint module. This data is further stored in cloud database.

Verification mode: In this verification mode the system verifies the student by scanning the fingerprint of the candidate and compares with the stored fingerprints data.

Scanning mode: If the fingerprint is registered, the LCD displays "CANDIDATE NAME and REGISTRATION NUMBER". This confirms the authenticity of the fingerprint. The Wi-Fi module sends notification to the cloud database that the registered candidate is present for the exam. When a wrong finger is placed on the module, the scanned image is sent to the fingerprint module. After browsing the database in its memory, if no match found, it will displayed on the screen a message "INVALID LOGIN ATTEMPT".

Storing mode: Further the details of successful login will be sent over the Cloud server which can be seen by the authority at the backend on any of the computer and the data will also be stored their which can be fetched anytime whenever needed.

5 Results and Discussions

The project is extremely easy to operate and can easily be modified according to future scopes. For the examination have use fingerprint biometric authentication. It is a security process that relies on the unique fingerprint of the candidate will be used to verify that the authenticity of the candidate appearing for the exam. Thus, the system inherently differentiates between an authorized person and a fraudulent imposter.

Figure 3 shows the hardware implementation of the authentication system The fingerprints of the candidates have been enrolled and store in the fingerprint module. Candidates are allowed 3 login attempts. When the candidate put his finger on the fingerprint module sensor, firstly sensor will enrolled the finger in the database of the module and stored this result in module and sensor will match this result which

Fig. 3 Hardware Implementation of the authentication system

is enrolled in this sensor with the help of fingerprints stored in cloud database and when it matches the result, the cloud server send command to the device to print Candidates name and registration number of the respective candidate. In case, the candidate fails to verification even after 3 login attempts, the cloud database will send command to the device to print "INVALID LOGIN ATTEMPT".

Figure 4 shows verification on Adriano ide serial monitor. It shows the start of verification.

Figure 5 shows results of authentication of some students on LCD Display ("CANDIDATE NAME" and "REGISTRATION NUMBER").

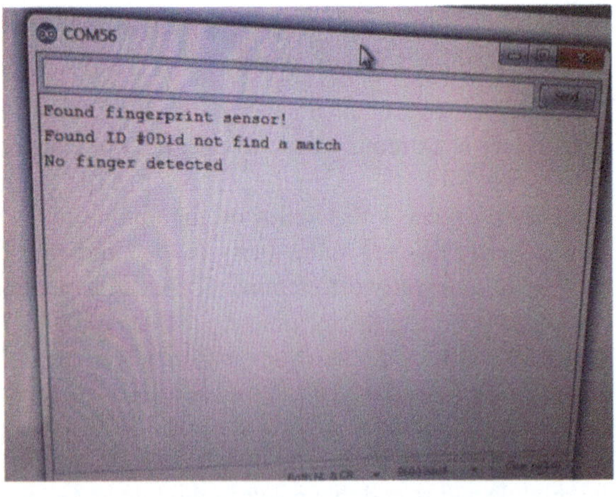

Fig. 4 Candidate verification on adruino ide serial monitor

Fig. 5 Results of authentication of students on LCD Display

Figure 6 shows results of authentication of some students on **a** webpage server ("CANDIDATE NAME","ROLL NO","REGISTRATION ID","TIME").

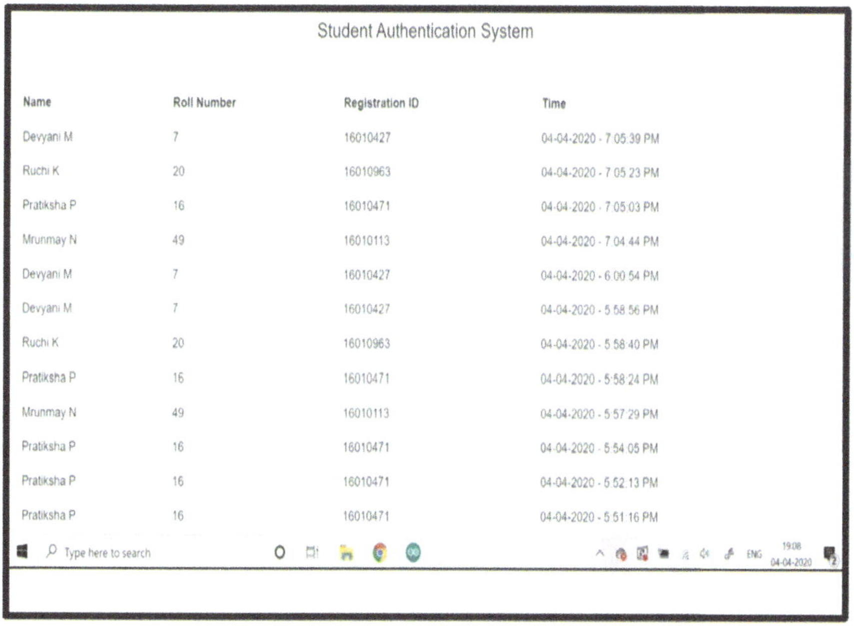

Student Authentication System

Name	Roll Number	Registration ID	Time
Devyani M	7	16010427	04-04-2020 - 7:05:39 PM
Ruchi K	20	16010963	04-04-2020 - 7:05:23 PM
Pratiksha P	16	16010471	04-04-2020 - 7:05:03 PM
Mrunmay N	49	16010113	04-04-2020 - 7:04:44 PM
Devyani M	7	16010427	04-04-2020 - 6:00:54 PM
Devyani M	7	16010427	04-04-2020 - 5:58:56 PM
Ruchi K	20	16010963	04-04-2020 - 5:58:40 PM
Pratiksha P	16	16010471	04-04-2020 - 5:58:24 PM
Mrunmay N	49	16010113	04-04-2020 - 5:57:29 PM
Pratiksha P	16	16010471	04-04-2020 - 5:54:05 PM
Pratiksha P	16	16010471	04-04-2020 - 5:52:13 PM
Pratiksha P	16	16010471	04-04-2020 - 5:51:16 PM

Fig. 6 Results on a webpage server

6 Conclusion

Authentication within the examination hall is one of the most important issue. To maintain attendance verification database is also a complicated and time-consuming process. Solution to these issues to the design fingerprint based exam hall authentication system that can pass only the candidates verified by their fingerprint scan and reject non verified candidates. The main objective is to reduce any form of imitation during exam by employing a more secured system based on fingerprint biometrics. In the proposed system is designed for authentication of a candidate or student at the time of examination. The experimental results shows that the method is suitable for all the authentication-based application and also its robustness makes it a useful system for implementation to specified areas. Further the details of successful login will be sent over the cloud server which can be accessed by the authority at the backend and the data will be stored for future access.

References

1. Ratha, N.K., Bolle, R.M., Pandit, V.D., Vaish, V.: Robust fingerprint authentication using local structural similarity. In: Proceedings Fifth IEEE Workshop on Applications of Computer Vision (2000)
2. Dhanalakshmi, N., Kumar, S.G., Padma Sai, Y.: Aadhaar based biometric attendance system using wireless fingerprint terminals. In: 2017 IEEE 7th International Advance Computing Conference (IACC) (2017)
3. Mohamed, B.K.P., Raghu, C.V.: Fingerprint attendance system for classroom needs. In: 2012 Annual IEEE India Conference (INDICON) (2012)
4. Kulkarni Vaishnavi, V., Waghmare Mangita, S., Gund Supriya, K., Shivpuje Dananjay, B.: Fingerprint based exam hall authentication. Int. J. Res. Eng. Sci. Manage. **2**(10) (2019)
5. Alsaadi, I.M.: Physiological biometric authentication systems, advantages, disadvantages and future development: a review. Int. J. Sci. Technol. Res. **4**(12), 285–289 (2015)
6. Alshbtat, A., Alfraheed, M., Zanoon, N.: A novel secure fingerprint-based authentication system for student's examination system. (IJACSA) Int. J. Adv. Comput. Sci. Appl. **10**(9) (2019)
7. Shrestha, A., Rakshit, T.: Experimental design and implementation of fingerprint based exam hall authentication system with temperature sensing and analysis using internet of things. In: International Conference on Interdisciplinary Cyber Physical Systems (ICPS) (2020)
8. Prince, N., Sengupta, A., Unni, K.: Implementation of IoT based attendance system on a dedicated web-server. Int. J. Sci. Eng. Res. **7**(6) (2016). ISSN 2229-5518
9. Win, Y.M., Nyein, S.A.: Wireless student attendance system using fingerprint sensor. Int. J. Trend Sci. Res. Develop. (IJTSRD) (2019)
10. Thein, M.M.M., Tun, C.M.N.H.M.: Students' attendance management system based on RFID and fingerprint reader. Int. J. Sci. Technol. Res. **4**(07) (2015)

Pandemic Surveillance Through Perspective Transformation Using YOLO and Mobile Net

Prachi Palsodkar, Prasanna Palsodkar, Yogita Dubey, and Roshan Umate

1 Introduction

In the year 2021, vaccination against coronavirus is completing fast and the world is trying to start with new normal. But the vaccination rate is not too high and it doesn't cover hundred percent assurances. To avoid the pandemic spread, social norms need to be followed. Many mutants have been observed which is another major concern that needs to be notified in public. The best preventive majors under such circumstances are to keep social distance and wear the mask in all crowded places like malls, markets, shops, schools, etc.

Considering the current scenario, tracking of social distancing is essential. A pandemic surveillance system is an approach that will be monitoring social distancing in specific areas like at the entrance of the building, mall, etc., and it will check whether the people present in the area are wearing masks or not. The system will be using CCTV feed to monitor the area. It uses object detection models like YOLO, Mobile-Net that help to detect the pedestrians on the frame of the image and then the system calculates the distance between them by performing the perspective transformation on the frame. The system also performs mask detection on the image to detect the mask on the pedestrians. The system considers these parameters to decide the output like the density of people, number of mask violations and number of social distancing violations [1]. The threshold limit on the density of people is set based on the government guidelines issued in the area. The system treats the social distancing

P. Palsodkar (✉) · P. Palsodkar · Y. Dubey
Yeshwantrao Chavan College of Engineering, Nagpur, Maharashtra, India
e-mail: prachi.palsodkar@gmail.com

R. Umate
Jawaharlal Nehru Medical College, Datta Meghe Institute of Medical Sciences (DU), Wardha, India

S. Mukherjee et al. (eds.), *Intelligent Systems for Social Good*,
Advanced Technologies and Societal Change,
https://doi.org/10.1007/978-981-19-0770-8_16

violation if the distance between two people is less than 6 feet. The alert signal by the system is given as one of the three, such as low risk, medium and high risk [2, 3].

The main objective in this chapter is to check the pedestrian flow and predict the distance with a warning alert system. This chapter will cover related work in the same area in Sect. 2, Preliminary Methodologies in Sect. 3, experimentation in Sect. 4 and result and conclusion in subsequent sections.

2 Related Work

COVID-19 spreads mainly among people who are in close contact (within about 6 feet) for a lengthened period. Spread happens when an infected person coughs, sneezes, or talks, and droplets from their mouth or nose float into the air and land in the mouth or nose of people nearby [2]. We need social distancing devices to ensure social distancing without any human interaction. Mostly there are wearable bands or watches that use proximity sensors that detect any nearby object, but it is not a protected way to maintain social distancing as it increases the risk of spreading COVID-19 through the band itself [4].

During the pandemic, significant work has been carried out on social distancing detection, Yang et al. [5] proposed a warning system using region-of-interest. Foster R-CNN model and YOLO4 was used to detect the real-time behaviour of the pedestrian. Zhou et al. [6] used MDA approach to check the crowd behaviour. They have used crowd collective behaviour classification for behaviour prediction. Punn et al. used the deep learning YOLO v3 technique applicable to various situations and used it to monitor real-time pedestrians. Real-time voice alert for workers' safety is discussed by Khandelwal et al. [3]. A similar kind of safety assist is discussed by Bochkovskiy et al. [7] using YOLO v4 model. Visual social distancing in surveillance also creates a privacy problem for individuals [8].

Many popular object detection models are available like Le-Net, Res-Net, Mobile Net, R-CNN, YOLO. The selection of an appropriate model was the challenge for the said experimentation. It was carried out with the following study.

Mobile Net V2 and SSD Lite show better accuracy with a lower number of parameters. Also, it is good for low computational devices with low Madd value. YOLO v3 [9] is comparatively very fast and accurate. It has faster FPS (frame Rate) with less background errors. With reference to Fig. 1 YOLO model is selected for the analysis.

3 Methodology

In "Pandemic surveillance system" the whole system is divided into three stages, namely the Input stage, Processing stage, and Alert stage as shown in Fig. 2.

The overview of each stage is as follows:

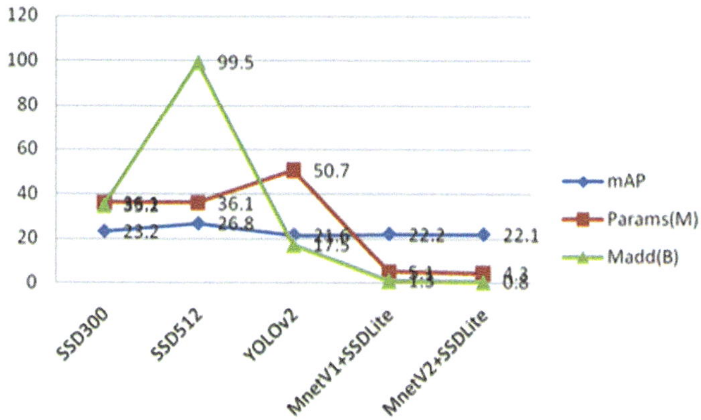

Fig. 1 Comparison of object detection models based on mAP, Params, and Madd

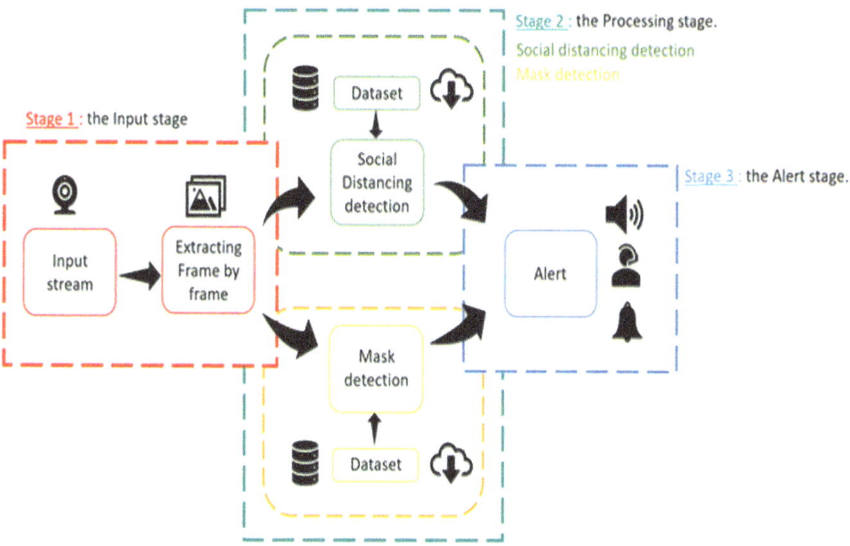

Fig. 2 System overview

Input stage

This stage is just a data transformation stage of the system which takes video from the CCTV continuously as input. This input video sequence will be converted into frames. These frames are later fed to the processing stage.

Processing stage: This stage contains two substages which will be executed simultaneously:

- **Social distancing detection**: This is one of the substages of the second phase of the project, which will run parallel with the Mask Detection stage. In this phase, the frame extracted from the input stream will be passed through the object detection model to detect the pedestrian in the frame and the coordinates of the detected pedestrian will be used to calculate the distance between them. The distance measurement is not simple as the camera is in a fixed position so it perceives the object's distance incorrectly; hence we use perspective transformation also called Bird's Eye View transformation. Now, after the transformation, the Euclidean distance will be calculated between the detected pedestrian and then it will be compared to the threshold (mostly 6 feet) and if the distance is less than that threshold, it will send a signal to the alert block of the project.

- **Mask detection**: This substage will receive input from the input stage. This stage will check whether a person is wearing a mask or not. Here we will have a face mask detection model based on computer vision and deep learning. If anybody is found not wearing a mask, this substage will provide this feedback to the output stage.

Alert stage: As we have seen in the processing stage, the social distancing violation data and mask violation data will be fed to this stage. This stage will then, according to the input from the processing stage, check whether there is any violation or not and also the level of severity of that violation. This will be followed by taking appropriate steps of alert. In case of high severity, it will also inform the security authority.

Working of the system consists of the input stage, as per the requirement, the area to be monitored is marked as the Region of Interest (ROI) by marking 4 points in a predefined order. Set the threshold distance for the minimum distance between two humans. After this, the working of the second stage starts, i.e. processing stage. The Processing stage has two subparts running parallel. One of them is Social Distancing Detection; in this stage, the model extracts the features from the frame of the image. The model uses CNN layers with YOLOv4 backbone which is already trained on the COCO dataset. The training gives the value of weights for the model to give accurate and efficient results. By using the YOLOv4 backbone, the model has extracted features with the probabilities of object (present in COCO dataset) in the given ROI. The model also localizes the feature by keeping a track of the coordinate of the feature on the frame of the image. These coordinates are used to perform Bird Eye View Transformation. The coordinates refer to the bounding box around the Pedestrian detected, calculate the Euclidean distance between the bounding boxes by first performing Birds Eye View Transformation to the frame of the image. The distance is compared to the threshold to calculate the social distancing violation whose count is passed to the alert stage.

Step by step detection process is discussed in this section, which is as follows:

3.1 Pedestrian Detection

In this stage, examine pedestrian detection. The input feed from the input stage will be given to the pedestrian detection stage using the YOLOv4 framework which uses CNN to perform pedestrian detection. In the YOLOv4 framework after input, the backbone does the feature extraction process with the help of CNN. Here each layer produces a feature map with the help of a convolution layer and pooling layer. After this, the layers present at the neck section collect all the feature maps generated at different layers of the backbone. To detect pedestrians, weights are used which are obtained by training the model on the COCO dataset which has a person class. Figure 3 shows the process of Pedestrian detection.

- Object Localization: The backbone network has the location of different features extracted in the form of a feature map. This helps the system to know the location of a feature extracted. These are calculated with the help of kernels [10].
- Image Classification: After the features are extracted, the feature map is fed to the three different heads after passing through Feature Pyramid Network (FPN). The FPN is used because it enables the framework to learn objects of three different sizes. The three different heads use three different anchor boxes, i.e. (19 × 19), (38 × 38), (76 × 76). Each head predicts the following parameters: Box centre (X, Y), Box size, Object-ness score, and Class score. The total confidence score

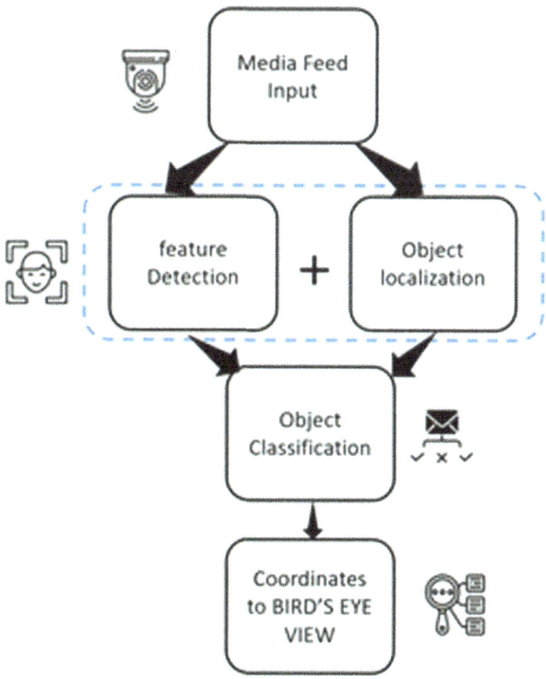

Fig. 3 Workflow for object detection

for a class is a product of object-ness score and class score.

$$\text{Confidence Score} = \text{Object Score} * \text{Class Score} \quad (1)$$

The coordinates of the pedestrian, i.e. box centre are now sent to the next stage which is Bird's Eye Transformation.

3.2 *Mask Detection*

Data set is a collection of millions of images used to train the computer. We need a very large data set of people wearing face masks and people not wearing face masks [11].

3.3 *Bird's Eye View*

Custom trained YOLOv4 model used to detect a person on each frame. The output from the model is a list of coordinates of bounding rectangles on detected persons; where a single rectangle is represented as: [x-min, y-min, width, height]. Social Distance is determined by calculating the Euclidean distance between the bounding rectangles. The imaging plane or the camera plane is a 2D projection of 3D world coordinates, therefore, the spatial relationship between the objects in this plane changes due to camera perspective. The objects near to the camera appear larger than those that are away from it. Calculating the distance between the rectangles in this perspective would give an imprecise estimate of the actual distance, we must correct it by transforming the image into top-down-view or bird's-eye-view. We have already discussed how this transformation works in the previous section. After the perspective transformation is done, we can now calculate the Euclidean distance between objects (people in our case) with more precision [12]. Figure 4 shows a flow for Bird's Eye View.

Then, Euclidean distance (ED) for real time distance is calculated as

$$\text{RTEC} = (6/\text{ED per feet}) * \text{Real time distance} \quad (2)$$

where RTEC stands for Real time Euclidean distance.

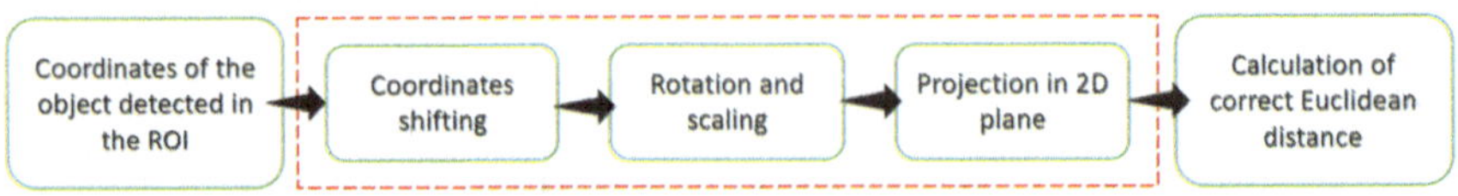

Fig. 4 Bird's eye view

Table 1 Level of violation

Parameters level	Social distancing violation	Mask violation	Number of people
Safe	No	No	Less than threshold value
Low risk	Yes	No	A little above threshold
High risk	Yes	Yes	Above threshold

As shown in Eq. (2), if RTEC is below 6 feet, then an alert is created.

3.4 Alert

This is the last stage which receives input from the mask detection and bird's eye view stage. The bird's eye view stage calculates the distance which will be fed to the alert stage which decides whether the distance is safe or not. Also, the total number of people in the ROI will be fed within that particular time.

As per Table 1, the inputs received, the alert stage decides the level of violation. These levels of violation are based on the social distancing violation, number of people and mask violation. The threshold value will be determined according to the size of ROI. According to the level of violation, it will alert the authority.

4 Results and Discussion

The process begins with the CCTV feed fed into an object detection model after converting it into frames. We are using the YOLOv4 model for object detection. Along with the object detected in the bounding box, it also returns the coordinates of the objects detected which will be fed to the perspective transformation stage for further processing [13].

Figure 8 shows those coordinates that we get after the object detection as shown in Fig. 7 when Fig. 6 is provided as input. Those coordinates will be used for the calculation of Euclidean distance for the detection of social distancing violations (Fig. 8).

The mask detection model works parallel with the people detection part and takes the CCTV feed. This part returns whether the person in the image is wearing a mask or not as shown in Fig. 5. This information will be important for deciding the level of risk in the alert stage [15]. For experimentation purpose, 1315 data points were

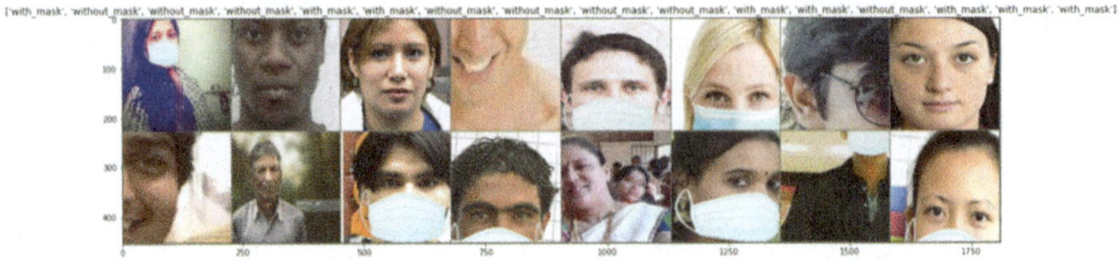

Fig. 5 Mask detection (Dataset source [14])

Fig. 6 Input image for object detection

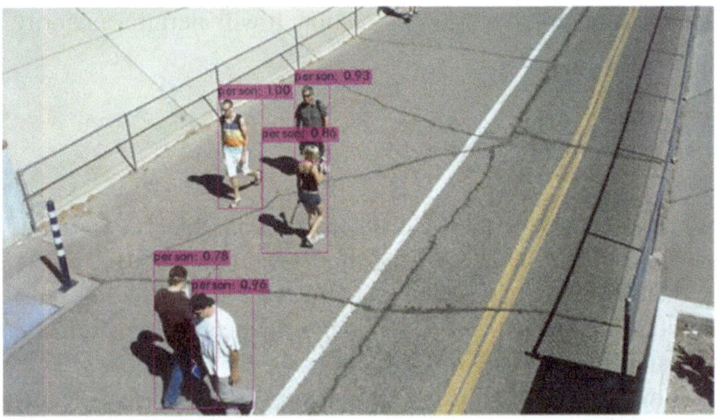

Fig. 7 Output image for object detection

considered for testing and 194 data points were considered for training (Figs. 9 and 10).

The coordinates and the image are fed to the perspective transformation stage for precise calculation of the Euclidean distance. The output will show the top view of

```
data/bev2.JPG: Predicted in 53.946000 milli-seconds.
person: 78%     (left_x:  255   top_y:  416   width:  103   height:  232)
person: 96%     (left_x:  316   top_y:  463   width:  102   height:  185)
person: 100%    (left_x:  361   top_y:  149   width:   73   height:  176)
person: 86%     (left_x:  433   top_y:  218   width:  108   height:  179)
person: 93%     (left_x:  486   top_y:  127   width:   58   height:  124)
person: 48%     (left_x:  488   top_y:  225   width:   55   height:  165)
person: 49%     (left_x:  624   top_y:   -0   width:   63   height:   30)
```

Fig. 8 Coordinates of detected people after detection

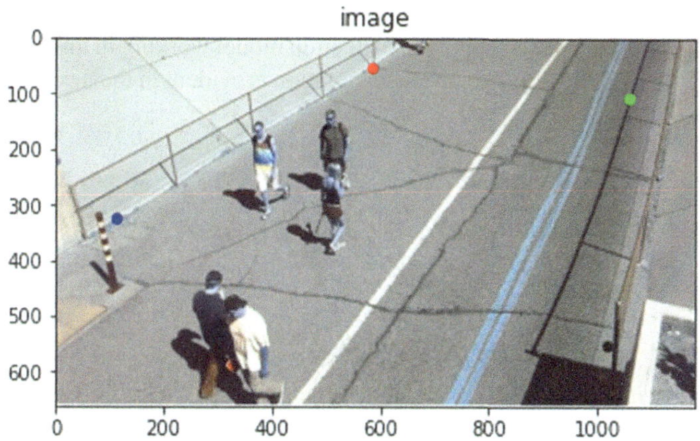

Fig. 9 Input images for perspective transformation

Fig. 10 Output perspective view of the image

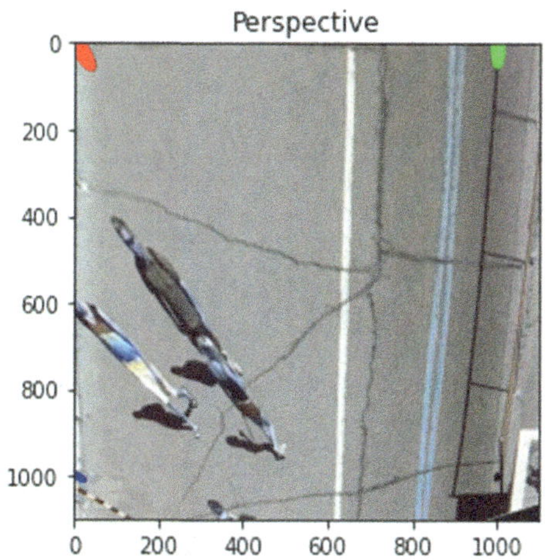

the region of interest. This calculation will allow us to check for social distancing violations in the alert stage.

Figure 11 shows the top view of the CCTV frame running in the background. Whenever there is any person detected in the frame, it is represented as a dot as shown in Fig. 10. The colour of the line between the dots is green whenever those people are at a safe distance from the other people (dot) in the surroundings. Otherwise, it is shown with a red line between the dots.

Figure 11 shows 4 points along the corner represents the ROI and the two points in the middle represents 6 feet distance required to maintain social distancing which is set by health organizations worldwide for safer interaction.

Figure 12 shows the ROI and object detection model working in unison. People in the ROI are being detected. The distance violation is evident in the next image which

Fig. 11 Coordinates showing ROI and the middle coordinates to set the limit

Fig. 12 Peripheral showing ROI

```
Processing frame:  1
[[   0.           629.6864299   740.46269319 1395.33401019]
 [ 629.6864299     0.          111.3283432   785.39798828]
 [ 740.46269319  111.3283432     0.          686.56827774]
 [1395.33401019  785.39798828  686.56827774    0.        ]]
Processing frame:  2
[[  0.         638.11127556 738.76450916]
 [638.11127556   0.         100.66280346]
 [738.76450916 100.66280346   0.        ]]
Processing frame:  3
[[   0.           648.54915003 1376.1809474   744.8288394 ]
 [ 648.54915003    0.          751.28822698   96.30160954]
 [1376.1809474   751.28822698    0.          663.31591267]
 [ 744.8288394    96.30160954  663.31591267    0.        ]]
Processing frame:  4
[[   0.           649.52059244 1367.18872143  740.21888655]
 [ 649.52059244    0.          741.37305049   90.80198236]
 [1367.18872143  741.37305049    0.          659.57031467]
 [ 740.21888655   90.80198236  659.57031467    0.        ]]
```

Fig. 13 Outputs coordinates values

shows a red line between the people when they are walking within 6 feet range and hence violating the norm.

Figure 13 shows output coordinate values for frames.

The distance matrix is calculated at each frame as shown in Fig. 13, these matrices show the distance of every person from every other person. Figure 14 shows a bird's eye view showing the distance between two persons.

For example, in the first frame, 4 pedestrians are identified and the distance matrix is a 4×4 matrix showing their distances from each other.

There are two objectives of object detection models like YOLO, R-CNN, etc. One is classification and the second is localization. In order to evaluate the performance of a model, we use the idea of IOU (Intersection over Union), which is the ratio of intersection of the region of ground truth and predicted bounding boxes to the union of the region of ground truth and predicted boxes. If the IOU is 1, it means that the ground truth and predicted boxes overlap and the detection is correct. Image size for YOLO v4 considered is Width = 512, Height = 512.

Table 2 shows the Precision, recall, and F1 score, appropriate selection of threshold plays a major role in distance estimation.

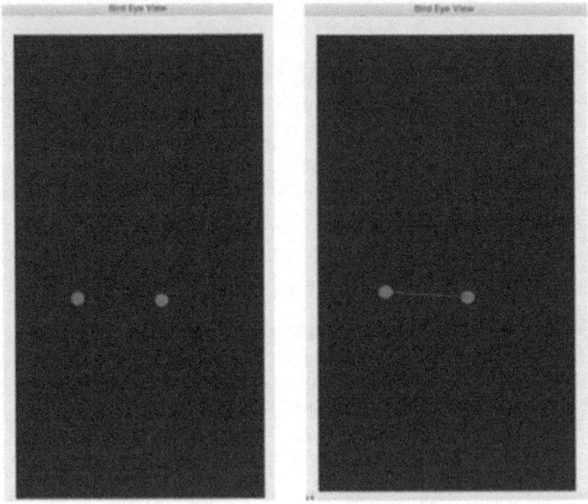

Fig. 14 The bird's eye view showing distance between two persons

Table 2 Precision, recall, and F1 score

No. of images (training)	Threshold (classes)	Precision	Recall	F1 score
608	25	0.66	0.73	0.68
608	80	0.90	0.42	0.54
320	25	0.64	0.60	0.61
320	80	0.89	0.60	0.61

5 Conclusion

This chapter gives a view of computer vision-based surveillance during pandemic using a perspective transformation. Perspective transformation gives more accurate distance calculations, thus the chances of the accuracy of the system in the detection increases. The methodology also suggests the use of cloud technology not only for the places with no computational power, but also for all places owing to the advantage of low cost and pay-as-you-go model in the use of cloud technology making the implementation cost-effective. Object detection and deep learning models uses birds eye view and YOLO v4 technique. Bounding boxes determines the actual distance and trigger the alert system.

Future work requires more generalisation of the algorithm where all possible angles of face will be covered for mask detection and the density of the crowd will be same as Indian market population density.

References

1. Amin, P.N., Moghe, S.S., Prabhakar, S.N., Nehete, C.M.: Deep learning based face mask detection and crowd counting. In: 2021 6th International Conference for Convergence in Technology (I2CT), 2021, pp 1–5 (2021). https://doi.org/10.1109/I2CT51068.2021.9417826
2. Punn, N.S., Sonbhadra, S.K., Agarwal, S.: Monitoring COVID-19 social distancing with person detection and tracking via fine-tuned YOLOv3 and Deepsort techniques. arXiv:2005.01385 (2020)
3. Khandelwal, P., Khandelwal, A., Agarwal, S., Thomas, D., Xavier, N., Raghuraman, A.: Using computer vision to enhance safety of workforce in manufacturing in a post COVID world arXiv:2005.05287 (2020)
4. Nguyen, C.T., Saputra, Y.M., Van Huynh, N., Nguyen, N.T., Khoa, T.V., Tuan, B.M., Nguyen, D.N., Hoang, D.T., Vu, T.X., Dutkiewicz, E., et al.: Enabling and Emerging Technologies for Social Distancing: A Comprehensive Survey. arXiv 2020. arXiv:2005.02816 (2005)
5. Yang, D., Yurtsever, E., Renganathan, V., Redmill, K.A., Özgüner, Ü.: A vision-based social distancing and critical density detection system for COVID-19. Sensors **21**, 4608 (2021) (2020). https://doi.org/10.3390/s21134608
6. Zhou, B., Wang, X., Tang, X.: Understanding collective crowd behaviors: Learning a mixture model of dynamic pedestrian-agents. In: Proceedings of the 2012 IEEE Conference on Computer Vision and Pattern Recognition, Providence, RI, USA, 16–21 June 2012, pp. 2871–2878 (2012)
7. Bochkovskiy, A., Wang, C.-Y., Liao, H.-Y.M.: YOLOv4: optimal speed and accuracy of object detection. arXiv:2004.10934 (2020)
8. Cristani, M., Del Bue, A., Murino, V., Setti, F., Vinciarelli, A.: The visual social distancing problem. IEEE Access **8**, 126876–126886 (2020)
9. Punn, N.S., Sonbhadra, S.K., Agarwal, S.: Monitoring COVID-19 social distancing with person detection and tracking via fine-tuned YOLO v3 and Deepsort techniques. arXiv 2020. arXiv: 2005.01385 (2020)
10. Zhao, Z.Q., Zheng, P., Xu, S.t., Wu, X.: Object detection with deep learning: a review. IEEE Trans. Neural Netw. Learn. Syst. **30**, 3212–3232 (2019)
11. Kumar, A., Kalia, A., Verma, K., Sharma, A., Kaushal, M.: Scaling up face masks detection with YOLO on a novel dataset. Optik **239**, 166744 (2021). ISSN 0030-4026. https://doi.org/10.1016/j.ijleo.2021.166744
12. Karaman, O., Alhudhaif, A., Polat, K.: Development of smart camera systems based on artificial intelligence network for social distance detection to fight against COVID-19. Appl. Soft Comput. **110**, 107610 (2021). ISSN 1568-4946. https://doi.org/10.1016/j.asoc.2021.107610
13. Zhao, Z.Q., Zheng, P., Xu, S.t., Wu, X.: Object detection with deep learning: a review. IEEE Trans. Neural Netw. Learn. Syst. **30**(11), 3212–3232 (2019)
14. Dataset. https://github.com/prajnasb/observations
15. Chowdary, G., Punn, N., Sonbhadra, S., Agarwal, S.: Face mask detection using transfer learning of inception V3 (2020). https://arxiv.org/2009.08369

The manufacturer's authorised representative in the EU is Springer Nature Customer Service Centre GmbH, Europaplatz 3, 69115 Heidelberg, Germany. If you have any concerns regarding our products, please contact ProductSafety@springernature.com

Printed and bound by CPI Group (UK) Ltd, Croydon, CR0 4YY
30/09/2026
02225245-0003